An Introduction to Stochastic Orders

An Introduction to Stochastic Orders

Félix Belzunce
Universidad de Murcia, Spain

Carolina Martínez-Riquelme
Universidad de Murcia, Spain

Julio Mulero
Universidad de Alicante, Spain

AMSTERDAM • BOSTON • HEIDELBERG • LONDON
NEW YORK • OXFORD • PARIS • SAN DIEGO
SAN FRANCISCO • SINGAPORE • SYDNEY • TOKYO
Academic Press is an imprint of Elsevier

Academic Press is an imprint of Elsevier
125 London Wall, London, EC2Y 5AS, UK
525 B Street, Suite 1800, San Diego, CA 92101-4495, USA
225 Wyman Street, Waltham, MA 02451, USA
The Boulevard, Langford Lane, Kidlington, Oxford OX5 1GB, UK

Notices
Knowledge and best practice in this field are constantly changing. As new research and experience
broaden our understanding, changes in research methods, professional practices, or medical treatment
may become necessary.

Practitioners and researchers must always rely on their own experience and knowledge in evaluating
and using any information, methods, compounds, or experiments described herein. In using such
information or methods they should be mindful of their own safety and the safety of others, including
parties for whom they have a professional responsibility.

To the fullest extent of the law, neither the Publisher nor the authors, contributors, or editors, assume
any liability for any injury and/or damage to persons or property as a matter of products liability,
negligence or otherwise, or from any use or operation of any methods, products, instructions, or ideas
contained in the material herein.

ISBN: 978-0-12-803768-3

British Library Cataloguing in Publication Data
A catalogue record for this book is available from the British Library

Library of Congress Cataloging-in-Publication Data
A catalog record for this book is available from the Library of Congress

For information on all Academic Press publications
visit our website at http://store.elsevier.com/

Working together
to grow libraries in
developing countries

www.elsevier.com • www.bookaid.org

To the loving memory of my parents.
Luis and *Magdalena*

F.B.

To my parents.
Antonio and *Carolina*

C.M.R.

To my family.

J.M.

CONTENTS

LIST OF FIGURES

LIST OF TABLES

As the title reveals, this book is an introduction to the topic of stochastic orders or, to be more precise, to the comparison of random quantities in a probabilistic sense, and it is oriented for graduate and PhD students. Stochastic orders are a powerful tool for the comparison of probabilistic models in different areas as reliability, survival analysis, risks, finance, and economics. The aim of this work is to provide a general background on this topic for students and researchers who want to use stochastic orders as a tool for their research.

The main reference on the topic is the monograph by Shaked and Shanthikumar [1], where, apart from providing the main results on the topic, there are some additional chapters on specific applications of stochastic orders in several fields. Shaked and Shanthikumar [2] is an updated version of their edition from 1994, without additional chapters on applications. These two books are considered a must for those interested in the topic. Of course there are some other books on the topic. Another important reference is the book Müller and Stoyan [3], where it is possible to find the main results on stochastic orders and several chapters on applications. Additional books devoted to stochastic orders are Stoyan [4], Kaas et al. [5], Denuit et al. [6], Levy [7], and Sriboonchitta et al. [8].

From our teaching experience, the books by Shaked and Shanthikumar [1, 2] are hard to read for graduate students or non-specialist researchers, because the discussion is mainly theoretical rather than applied. In this work, we have tried to avoid this problem providing applications of the main results to different probabilistic models of interest in several fields. Furthermore, we provide detailed proofs for most of the main results. Therefore, the readership can find in a monograph both detailed discussions of fundamental properties of several stochastic orders, in the univariate and multivariate cases, and applications to several probabilistic models.

The organization of the book is as follows. Chapter 1 is devoted to the introduction of several concepts for univariate and multivariate distributions. The definition of the univariate stochastic orders considered in this manuscript are given in terms of the comparison of several functions associated to univariate distributions. In this chapter, we provide the definition and interpretation of these functions in several contexts, like reliability, survival analysis, risks, and economics. In order to provide examples based on some data sets, we also give non-parametric estimators of such functions. In addition, we provide the definition of some parametric models of univariate distributions that will be used along the book to illustrate some results. Some additional results of total positivity theory are also provided. These results are commonly used in the proofs of some theorems and are a must for anyone interested in the topic. We also present some functions associated to multivariate distributions, as well as some parametric models of multivariate distributions and some dependence notions.

In Chapter 2, we start with the topic of this book in the univariate case. Here, we present the main stochastic orders considered in the literature for univariate distributions. In fact, in the literature, we can find a great number of stochastic orders, but here we have included only those that we have considered as the most used and attractive from an applied point of view. For these stochastic orders, we give the main results, which include characterizations, sufficient conditions for the stochastic orders to hold, and preservation under convergence, mixtures, transformations and convolutions. In order to illustrate the comparison with some real data sets, under the different criteria, of two populations, we provide as a preliminary tool some non-parametric estimators of the functions involved in the corresponding criteria. This is not intended as a statistical procedure for the validation of the different stochastic orders, and some statistical tests should be considered for such validation. This topic is not considered in this book and it would be interesting to see a specific monograph on the topic. This chapter ends with a section on applications, which includes several tables with a summary of conditions for the stochastic comparison of several parametric univariate distributions in the continuous and discrete cases. We also include a section on the comparison of coherent systems and distorted distributions, and individual and collective risks models.

Finally, in Chapter 3, we give an introduction to the topic of multivariate stochastic orders. Given that this material is not easy to handle for graduate students or non-experts on the topic, we have included what we have considered is the most relevant material. A section with applications to conditionally independent models and ordered data is also included.

Finally we would like to dedicate a few words to Moshe Shaked, who recently passed away. Moshe has been one of the leading contributors in the topic of stochastic orders. His influence is extremely broad and he has made fundamental contributions in any area related to stochastic orders. We shall miss Moshe and his wife Edith during upcoming conferences and events related to stochastic orders.

Félix Belzunce
Carolina Martínez-Riquelme
Julio Mulero

CHAPTER 1

Preliminaries

1.1 INTRODUCTION

As it was mentioned in the Preface, this book is oriented for graduate and PhD students and researchers which are interested in acquiring a basic understanding on the theory of stochastic orders. Hence, this book is intended as an introduction on this topic, and the reader is assumed to have a basic knowledge on probability theory and real analysis. Since basic courses on probability theory do not require measure theory notions, we have avoided any discussion where such notions are involved. In this chapter, several notions and results for univariate and multivariate distributions are recalled. In particular, the notions and results for univariate distributions will be given in Section 1.2 and the ones for multivariate distributions will be given in Section 1.3.

The following notation will be used along this book. Given any random variable or random vector X and any event A, $[X|A]$ denotes the conditional random variable X given the event A. The indicator function for a set A will be denoted by I_A, that is, $I_A(x) = 1$, if $x \in A$ and $I_A(x) = 0$, if $x \notin A$. Finally, the equality in law will be denoted by $=_{\mathrm{st}}$.

1.2 UNIVARIATE DISTRIBUTION NOTIONS

In this section, the main notions and results related to univariate distributions, which will be used in the book, are provided. In general, the definitions are related to the distribution function of a random variable.

Given a *random variable X* defined on a probability space, it is defined its *distribution function*, denoted by F, as

$$F(x) = P[X \leq x], \quad \text{for all } x \in \mathbb{R}.$$

If the distribution function can be represented as

$$F(x) = \int_{-\infty}^{x} f(t)\, dt,$$

An Introduction to Stochastic Orders. http://dx.doi.org/10.1016/B978-0-12-803768-3.00001-6

1

where f is known as the *density function*, we will say that X is *continuous* (the term absolutely continuous is also used in probability theory). In such case, the support of X is a union of intervals. If the support of X is a discrete set on $\mathbb{R}$, then we will say that X is *discrete* and, in this case, it is defined its *mass probability function*, denoted by p, as

$$p(x) = P[X = x], \quad \text{for all } x \text{ in the support of } X.$$

The left and right extremes of the support of a random variable X will be denoted by l_X and u_X, respectively.

Let us see some functions related to a distribution function through which will be defined the different criteria to compare random variables. Applications of these functions in reliability, survival analysis, risks, and economics will be also described.

1.2.1 The survival and the quantile function

Two main functions related to a distribution function are the survival and quantile functions. Let us see the formal definitions.

Given a random variable X with distribution function F, it is defined its *survival function*, denoted by $\overline{F}$, as

$$\overline{F}(x) = 1 - F(x), \quad \text{for all } x \in \mathbb{R}.$$

In reliability and survival analysis, where a random variable represents the random lifetime of a unit, a device or an organism, the survival function is the probability to survive beyond x units of time. Obviously, the survival function is a decreasing function, and $\lim_{x \to -\infty} \overline{F}(x) = 1$ and $\lim_{x \to +\infty} \overline{F}(x) = 0$. If X is non-negative with finite mean, then

$$E[X] = \int_0^{+\infty} \overline{F}(x)\, dx. \tag{1.1}$$

Besides the survival function, it is defined the *quantile function* of X, denoted by F^{-1}, as

$$F^{-1}(p) = \inf\{x \in \mathbb{R} | F(x) \geq p\}, \quad \text{for all } p \in (0,1),$$

and it is the lower value such that the distribution function exceeds the value p. Some properties of the quantile function that will be used later are the following:

(i) $\{x \in \mathbb{R} | F(x) \geq p\} = [F^{-1}(p), +\infty)$.

(ii) Given a random variable U uniformly distributed on the interval $(0, 1)$, then $F^{-1}(U) =_{\text{st}} X$. Also recall that $U =_{\text{st}} F(X)$.

(iii) $p \leq F(F^{-1}(p))$ and the equality holds if $p = F(x)$, for some $x \in \mathbb{R}$.

(iv) If $p = F(x)$, then $F^{-1}(p) \leq x$.

(v) For any real valued increasing function h, the quantile function of $h(X)$ is given by $h(F^{-1}(p))$.

(vi) For a random variable X with finite mean, it holds that

$$E[X] = \int_0^1 F^{-1}(p)\, \mathrm{d}p. \tag{1.2}$$

It is also possible to define the generalized inverse of an increasing function ϕ, not necessarily strictly increasing, as

$$\phi^{-1}(t) = \inf\{x \in \mathbb{R} | \phi(x) \geq t\}, \quad \text{for all } t \in \mathbb{R}.$$

To illustrate with real data sets the different criteria that we are going to consider in Chapter 2, it is necessary to introduce the non-parametric estimators of the survival and quantile functions. Let us see the definition of such estimators.

Given a random sample $X_1, \ldots, X_n$ of a random variable X, it is defined its *empirical distribution function*, denoted by F_n, as

$$F_n(x) = \frac{1}{n}\sum_{i=1}^n I_{(-\infty, x]}(X_i), \quad \text{for all } x \in \mathbb{R}.$$

Then, it is defined its *empirical survival function*, denoted by $\overline{F}_n$, as

$$\overline{F}_n(x) = 1 - F_n(x), \quad \text{for all } x \in \mathbb{R}.$$

The Glivenko–Cantelli's theorem shows the uniform convergence of $\overline{F}_n$ to $\overline{F}$. Denoting by $X_{1:n} \leq \cdots \leq X_{n:n}$ the ordered sample, the non-parametric estimation of the quantile function is based on one or two order statistics. Here we consider the default non-parametric estimators given by the statistical package R. In this case, following Hyndman and Fan [9], the non-parametric estimators are given by

$$Q_n(p) = (1 - \gamma)X_{j:n} + \gamma X_{j+1:n},$$

where $(j - m)/n = p < (j - m + 1)/n$, and the values of γ and m depends on the sample quantile type. In particular, Hyndman and Fan [9] consider nine types and the default method is Type 7.

1.2.2 The stop-loss function

Given a random variable X, its *stop-loss function* is defined as

$$E[(X - x)_+], \quad \text{for all } x \in \mathbb{R},$$

where $(x)_+ = x$, if $x \geq 0$ and $(x)_+ = 0$, if $x < 0$. The stop-loss function is well known in the context of actuarial risks. If the random variable X denotes the random risk for an insurance company, it is very common that the company pass on parts of it to a reinsurance company. In particular, the first company bears the whole risk, as long as it is less than a fixed value x (called retention) and the reinsurance company will take over the amount $X - x$, if $X > x$. This is called a stop-loss contract with fixed retention x. From the point of view of the insurance company, the insurer handles $\min\{X, x\}$ and the reinsurer handles $(X - x)_+$. The expected cost for the reinsurance company is called the net premium, which is equal to $E[(X - x)_+]$. From (1.1), the stop-loss transform can be written in terms of the survival function $\overline{F}$ as

$$E[(X - x)_+] = \int_x^{+\infty} \overline{F}(t) \, dt. \tag{1.3}$$

Recall also that the existence of the stop-loss transform is ensured if the random variable has a finite mean.

Let us see a non-parametric estimator of the stop-loss function. The estimator can be constructed replacing the survival function by the empirical one in (1.3), that is, according to the previous notation, given $x \in [X_{i:n}, X_{i+1:n})$, for some $i \in \{1, \ldots, n - 1\}$, then *the empirical stop-loss function* is defined as

$$\int_x^{+\infty} \overline{F}_n(x) \, dx = (X_{i+1:n} - x)\frac{n - i}{n} + \sum_{j=i+2}^n (X_{j:n} - X_{j-1:n})\frac{n - j + 1}{n}.$$

If $x > X_{n:n}$, then

$$\int_x^{+\infty} \overline{F}_n(t) \, dt = 0,$$

and, if $x < X_{1:n}$, then

$$\int_x^{+\infty} \overline{F}_n(x) \, dx = \overline{X} + X_{1:n} - x,$$

where $\overline{X}$ denotes the sample mean.

1.2.3 The hazard rate and mean residual life functions

The hazard rate function has great interest in the reliability context. If the random variable X represents the lifetime of a unit or individual, this function measures the "probability" of instant failure at time x. Given a continuous random variable X with distribution function F and density function f, its *hazard rate function* is defined as

$$r(x) = \frac{f(x)}{\overline{F}(x)}, \quad \text{for all } x < u_X.$$

This function is related to the random variable $X_x = [X - x \mid X > x]$, which is known as *the residual lifetime* at time x. Since the survival function of X_x, denoted by $\overline{F}_x$, is given by

$$\overline{F}_x(t) = P[X - x > t \mid X > x] = \frac{\overline{F}(x + t)}{\overline{F}(x)}, \quad \text{for all } x \geq 0, \tag{1.4}$$

then

$$r(x) = \lim_{\Delta \to 0^+} \frac{1}{\Delta} \frac{\overline{F}(x + \Delta) - \overline{F}(x)}{\overline{F}(x)} = \lim_{\Delta \to 0^+} \frac{1}{\Delta} P[X_x \leq \Delta], \quad \text{for all } x \geq 0.$$

From this equality, it is obvious that this function is the intensity of failure of the unit. The property of having an increasing [decreasing] hazard rate function is called IFR [DFR], and it is a well-known and widely studied property. Equivalently, a random variable X is IFR [DFR] if, and only if, (1.4) is decreasing [increasing] in x, for all $t \geq 0$. The set of the random variables sharing this property is understood as an aging class in reliability theory. There exists a result on the preservation of this property under convolution. In particular, the following theorem is given by Barlow and Proschan [10].

Theorem 1.2.1. *Let X and Y be two independent random variables. If X and Y are IFR, then $X + Y$ is IFR.*

A related property to the IFR notion is that of log-concave density function or ILR notion. Given a continuous random variable X with density function f and interval support, we say that X is *increasing likelihood ratio*, denoted by ILR, if $\log f(x)$ is concave.

This property is also preserved under convolutions see Ref. [11].

Theorem 1.2.2. *Let X and Y be two independent continuous random variables with interval supports. If X and Y are ILR, then $X + Y$ is ILR.*

The relationship between the two previous notions is the following:

$$X \text{ is ILR} \implies X \text{ is IFR.} \tag{1.5}$$

There exists another function that is closely related to the hazard rate function: the mean residual life function. Given a random variable X with finite mean, its *mean residual life function* is defined as

$$m(x) = E[X_x], \quad \text{for all } x < u_X.$$

The mean residual life function measures the additional expected lifetime for a unit or individual which has survived up to time x. The mean residual life function can be interpreted in risk theory as well. In a stop-loss contract with fixed retention x, $m(x)$ measures the expected costs for the reinsurance company, from the point of view of the reinsurer. From (1.1) and (1.4), we have that

$$m(x) = \frac{\int_x^{+\infty} \overline{F}(x)\, dx}{\overline{F}(x)}, \quad \text{for all } x < u_X.$$

Analogously to the IFR notion, the property of having a decreasing [increasing] mean residual life function is called DMRL [IMRL]. Again, the set of the random variables sharing this property is understood as an aging class. The following relationship among the IFR and the DMRL notion holds:

$$X \text{ is IFR [DFR]} \implies X \text{ is DMRL [IMRL].}$$

1.2.4 Risk measures based on the quantile function

Given a random variable X with quantile function F^{-1}, it is defined the *value at risk*, denoted by VaR, as

$$\text{VaR}[X; p] = F^{-1}(p), \quad \text{for all } p \in (0, 1),$$

that is, it is the quantile function at point p. If the random variable is the risk associated to some action, like the potential loss in a portfolio position, then VaR$[X; p]$ is the larger risk for the $100p\%$ of the situations and gives the maximum risk, for a fixed time horizon, for the $100p\%$ of the cases. This is the main risk measure in risk theory. For a fixed p, the VaR does not provide information about the thickness of the upper tail of the distribution, thus some other measures have been considered for this purpose. These measures give information about the magnitude and variability of the losses beyond the value-at-risk.

Given a random risk X with distribution function F, besides the value at risk we have the following risk measures:

(i) *Tail value at risk*:

$$\text{TVaR}\,[X;p] = \frac{1}{1-p} \int_p^1 F^{-1}(u)\,\mathrm{d}u, \quad \text{for all } p \in (0,1).$$

The TVaR can be intuitively considered as the "arithmetic mean" of the VaRs of X from p on.

(ii) *Conditional tail expectation*:

$$\text{CTE}\,[X;p] = E\left[X\,\middle|\,X > F^{-1}(p)\right], \quad \text{for all } p \in (0,1).$$

It is the expected loss given that the loss exceeds its VaR.

(iii) *Conditional value at risk*:

$$\text{CVaR}\,[X;p] = E\left[X - F^{-1}(p)\,\middle|\,X > F^{-1}(p)\right] = m\left(F^{-1}(p)\right),$$

for all $p \in (0,1)$, where m denotes the mean residual life function of X. It is the additional expected loss to VaR given that the loss exceeds its VaR.

(iv) *Excess wealth or expected shortfall*:

$$
\begin{aligned}
W_X(p) &= E\left[\left(X - F^{-1}(p)\right)_+\right] \\
&= \int_{F^{-1}(p)}^{+\infty} \overline{F}(x)\,\mathrm{d}x \\
&= \int_p^1 (F^{-1}(q) - F^{-1}(p))\,\mathrm{d}q, \qquad (1.6)
\end{aligned}
$$

for all $p \in (0,1)$. The excess wealth measures the thickness of the upper tail from a fixed VaR, and it is also known as the *right spread function*.

When the random variable is continuous, we have the following equalities:

$$\text{TVaR}\,[X;p] = \text{CTE}\,[X;p]$$

and

$$W_X(p) = (1-p)\mathrm{CVaR}\,[X;p].$$

The excess wealth function is equal to the stop-loss function evaluated at $F^{-1}(p)$ and, therefore, it is the net premium for a stop-loss contract with fixed retention $x = F^{-1}(p)$.

Moreover, the excess wealth function can be interpreted also in the context of reliability. Burn-in testing is a common reliability technique to improve the lifetime of a product before selling. Let us consider a system which produces units with random lifetime X, where the units are tested until the $100p\%$ of the units fail (burn-in period), in order to eliminate early failures. From the point of view of the producer, the additional lifetime of the remaining $100(1-p)\%$ of the units is distributed as $(X - F^{-1}(p))_+$. Then, the expected lifetime of the units is $E[(X - F^{-1}(p))_+]$. Observe that, from the point of view of the consumer, the expected lifetime of the units is given by $\mathrm{CVaR}\,[X;p]$.

Notice that, for a non-negative random variable X, it holds that

$$E\left[\left(X - F^{-1}(p)\right)_+\right] = E[X] - \int_0^{F^{-1}(p)} \overline{F}(x)\,dx$$

and, according to Barlow et al. [12, pp. 235–237], given a random sample $X_1, \ldots, X_n$ of X ($F(0) = 0$), the non-parametric estimator of $E\left[(X - F^{-1}(p))_+\right]$ is defined as

$$W_n(p) = \overline{X} - H_n^{-1}(p), \quad \text{for all } 0 \leq p < \frac{1}{n},$$

where $\overline{X}$ is the sample mean, $H_n^{-1}(p) = nX_{(1)}p$, and

$$H_n^{-1}(p) = \frac{1}{n}\sum_{j=1}^{i}(n-j+1)(X_{j:n} - X_{j-1:n}) + \left(p - \frac{i}{n}\right)(n-i)(X_{i+1:n} - X_{i:n}),$$

for all $\frac{i}{n} \leq p < \frac{i+1}{n}$, where X_i denotes the ith order statistic of a sample of size n from X, and $X_0 = 0$.

1.2.5 Measures of concentration, inequality, and deprivation

In economy, measures of concentration and inequality arise in a natural way when the random variable represents the random annual family incomes. In this context, one of the most common measures is the Lorenz curve. Given a non-negative random variable X with quantile function F^{-1} and finite mean, it is defined its *Lorenz curve*, denoted by L_X, as

$$L_X(p) = \frac{1}{E[X]} \int_0^p F^{-1}(t)\, dt, \quad \text{for all } p \in (0,1).$$

Thus, $L_X(p)$ represents the share of total incomes received by the poorest $100p\%$ of the population. The Lorenz curve can be used to compute a well-known inequality measure, the Gini index, denoted by G_X. In particular, the Gini index can be computed as

$$G_X = 1 - 2 \int_0^1 L_X(p)\, dp.$$

Recently, Belzunce et al. [13] introduced a new inequality measure, the expected proportional shortfall, which can also be used to assess deprivation. Given a non-negative random variable X, it is defined the *expected proportional shortfall function*, denoted by EPS_X, as

$$\mathrm{EPS}_X(p) = E\left[\left(\frac{X - F^{-1}(p)}{F^{-1}(p)}\right)_+\right], \quad \text{for all } p \in (0,1).$$

This function can be defined for not necessarily non-negative random variables on the set $\{p \in (0,1)\,|\,F^{-1}(p) > 0\}$. However, we focus on the non-negative random variables.

In order to interpret this function as a measure of relative deprivation in the sense of Runciman [14], notice that the quantile function associated to the random variable

$$\left(\frac{X - F^{-1}(p)}{F^{-1}(p)}\right)_+$$

is given by

$$F_p^{-1}(q) = \begin{cases} 0 & \text{if } q < p, \\ \dfrac{F^{-1}(q) - F^{-1}(p)}{F^{-1}(p)} & \text{if } q \geq p. \end{cases} \tag{1.7}$$

According to (1.2) and (1.7), we get that

$$\mathrm{EPS}_X(p) = \frac{\int_p^1 \left(F^{-1}(q) - F^{-1}(p)\right)\, dq}{F^{-1}(p)}, \quad \text{for all } p \in (0,1).$$

Now, by aggregation, we see that the expected relative deprivation of an individual with rank p with respect to the whole population is given by

$\mathrm{EPS}_X(p)$, for all $p \in (0, 1)$. Thus, we can interpret $\mathrm{EPS}_X(p)$ as a measure of the relative deprivation felt by an individual with rank p in a population with distribution F.

Clearly, the expected proportional shortfall is scale invariant under any strictly positive scale change and, analogously to the case for the Lorenz curve, it does not depend on the base currency.

The expected proportional shortfall function is related to some of the measures considered in the previous section. In particular, we see that

$$\mathrm{EPS}_X(p) = \frac{W_X(p)}{\mathrm{VaR}[X; p]}$$

$$= (1 - p)\left[\frac{\mathrm{TVaR}[X; p]}{\mathrm{VaR}[X; p]} - 1\right] \qquad (1.8)$$

and, when the distribution function is continuous, we have the following equalities

$$\mathrm{EPS}_X(p) = (1 - p)\left[\frac{\mathrm{CTE}[X; p]}{\mathrm{VaR}[X; p]} - 1\right] \qquad (1.9)$$

$$= (1 - p)\frac{\mathrm{CVaR}[X; p]}{\mathrm{VaR}[X; p]}, \qquad (1.10)$$

for all $p \in (0, 1)$.

We can estimate the Lorenz curve by the empirical Lorenz curve, which is given by

$$L_n(p) = \frac{1}{\overline{X}}\int_0^p Q_n(q)\mathrm{d}q, \quad \text{for all } p \in (0, 1),$$

and the expected proportional shortfall can be estimated by

$$\mathrm{EPS}_n(p) = \frac{W_n(p)}{Q_n(p)}, \quad \text{for all } p \in (0, 1).$$

This expression arises by replacing the theoretical values in (1.8) with their empirical counterparts.

1.2.6 Total time on test transform

Given a non-negative random variable X with distribution function F, its *total time on test (ttt) transform* is defined as

$$\mathrm{ttt}_X(p) = \int_0^{F^{-1}(p)} \overline{F}(x)\,\mathrm{d}x, \quad \text{for all } p \in (0, 1).$$

Recently, Hu et al. [15] defined the following transform:

$$T_X(p) = E[X] - \int_{F^{-1}(p)}^{+\infty} \overline{F}(x)\, dx, \qquad \text{for all } p \in (0,1),$$

which extends the ttt transform for any random variable with finite mean. When X is non-negative, then $T_X(p)$ is reduced to the ttt transform, that is,

$$T_X(p) = \int_0^{F^{-1}(p)} \overline{F}(x)\, dx, \qquad \text{for all } p \in (0,1).$$

From the equality $\min\{x, t\} = x + (x - t)_+$, it is easy to see that

$$T_X(p) = E[X] - E[(X - F^{-1}(p))_+] = E[\min\{X, F^{-1}(p)\}]$$

and, therefore, the T_X transform can be interpreted in reliability and risk theory as follows.

According to the interpretation of the excess wealth function in reliability, considering again the context of burn-in testing, we see that T_X measures the expected lifetime of the units which have failed during the burn-in period.

In risk theory, for a stop-loss contract with fixed retention $x = F^{-1}(p)$, T_X is the expected cost for the insurer.

According to previous notation, if the random variable X is non-negative, the non-parametric estimator for T_X is given by $H_n^{-1}(p)$, for all $p \in (0,1)$.

1.2.7 Total positivity

In this section, a powerful tool to obtain results on the topic of stochastic orders is provided, which is a must for anyone interested in the theoretical part of the topic.

First, let us introduce the obtain of total positivity. Given $K : A \times B \subseteq \mathbb{R}^2 \mapsto \mathbb{R}$, we say that K is *totally positive of order* $r \in \mathbb{N}$, denoted by TP_r, if

$$\begin{vmatrix} K(x_1, y_1) & K(x_1, y_2) & \cdots & K(x_1, y_m) \\ K(x_2, y_1) & K(x_2, y_2) & \cdots & K(x_2, y_m) \\ \vdots & \vdots & & \vdots \\ K(x_m, y_1) & K(x_m, y_2) & \cdots & K(x_m, y_m) \end{vmatrix} \geq 0,$$

for all $x_1 < \cdots < x_m$ and $y_1 < \cdots < y_m$, where $x_i \in A$, $y_i \in B$, for all $i = 1, \ldots, m$ and $1 \le m \le r$.

From this notion, Karlin [16] gives one of the most well-known theorems in terms of this notion through the sign changes of a function, therefore a definition of this concept is required before stating the theorem. Given a function f, it is defined the sign changes of f on an interval $I \subseteq \mathbb{R}$, denoted by $S^-(f)$, as

$$S^-(f) = \sup\{S^-(f(x_1), \ldots, f(x_m))\},$$

where $S^-(f(x_1), \ldots, f(x_m))$ denotes the sign changes in this sequence and the supreme is taken among every ordered sequence in I, that means, for all $x_1 < \cdots < x_m$, for all $m \in \mathbb{N}$.

Let us see now the theorem given by Karlin [16]. This result is usually known as the variation diminishing property.

Theorem 1.2.3. *Let $K : A \times B \subseteq \mathbb{R}^2 \mapsto \mathbb{R}$ be a TP_r function, h a bounded and Borel measurable function on B and consider the transformation*

$$H(x) = \int_B K(x, y) h(y) \, dy.$$

If there exists $\int_B K(x, y) dy$, for all $x \in A$, and $S^-(h) \le r - 1$, then

$$S^-(H) \le S^-(h).$$

In particular, let X and Y be two continuous random variables with density functions f and g and distribution functions F and G, respectively. Through the previous theorem, it is possible to bound the sign changes of the difference of distribution functions, $S^-(G - F)$, by the sign changes of the difference of the density functions, $S^-(g - f)$. It is enough to consider $h(y) = g(y) - f(y)$ (which is obviously bounded) and $K(x, y) = I_{(-\infty, x]}(y)$ (it is easy to check that $K(x, y)$ is TP_r, for all $r \in \mathbb{N}$) and take into account that $\lim_{x \to 0}(G(x) - F(x)) = \lim_{x \to \infty}(G(x) - F(x)) = 0$. Then, we state the following result.

Corollary 1.2.4. *Let X and Y be two continuous random variables. If $S^-(g - f) \le n$, then*

$$S^-(G - F) < S^-(g - f).$$

There are a great number of results on TP_r functions. Next, we recall two of them which will be useful to prove several results of upcoming sections. The first one is known as the basic composition formula [16].

Theorem 1.2.5. *Let $K : A \times B \subseteq \mathbb{R}^2 \mapsto \mathbb{R}$ be TP_m, $L : B \times C \subseteq \mathbb{R}^2 \mapsto \mathbb{R}$ be TP_n and h be a bounded and Borel measurable function, then*

$$M(x, y) = \int K(x, z) L(z, y) h(z) \, dz$$

is $TP_{\min\{n,m\}}$ on $A \times C$.

The last theorem is given by Joagdev et al. [17].

Theorem 1.2.6. *Let h_1, h_2 be two Borel measurable real functions, such that h_1 is non-negative, h_2 may take negative values, and*

$$h_1(x)h_2(x') \geq h_1(x')h_2(x), \quad \text{for all } x \leq x' \text{ in } \mathbb{R}.$$

Further, suppose that $h_1(x)$ is increasing in x. Now, let $\overline{H}_i(x)$ be a survival function with support S such that $\overline{H}_i(x)$ is TP_2 in $\{1, 2\} \times S$, for $i = 1, 2$. Then, we see that

$$\int h_1(x) \, dH_1(x) \int h_2(x) \, dH_2(x) \geq \int h_1(x) \, dH_2(x) \int h_2(x) \, dH_1(x).$$

1.2.8 Parametric families of univariate distributions

In this section, some renowned parametric families are defined. These models will be used during the next chapter to aim a framework (as much complete as possible) on the comparison of these families in the several orders provided in the upcoming chapter. First, the continuous ones are defined. The definitions of the families are provided by means of the associated functions having explicit expressions. Next, we define the ones that only have closed expression for the density function, which is the case of the normal distribution.

Definition 1.2.7. Given a random variable X, it is said that X follows a *normal distribution* with mean $\mu \in \mathbb{R}$ and standard deviation $\sigma > 0$, denoted by $X \sim N(\mu, \sigma^2)$, if its density function is given by

$$f(x) = \frac{1}{\sigma \sqrt{2\pi}} \exp\left\{ -\frac{1}{2\sigma^2}(x - \mu)^2 \right\}, \quad \text{for all } x \in \mathbb{R}.$$

The normal distributions are ILR and, therefore, by (1.5), IFR as well.

A close family to the normal one is the well-known log-normal family, which can be obtained by the transformation $\exp\{X\}$, where $X \sim N(\mu, \sigma^2)$. Let us see the formal definition.

Definition 1.2.8. Given a random variable X, it is said that X follows a *log-normal distribution* with parameters $\mu \in \mathbb{R}$ and $\sigma > 0$, denoted by $X \sim LN(\mu, \sigma^2)$, if its density function is given by

$$f(x) = \frac{1}{x\sigma\sqrt{2\pi}} \exp\left\{-\frac{1}{2\sigma^2}(\log(x) - \mu)^2\right\}, \quad \text{for all } x \in (0, +\infty),$$

and its mean is given by $E[X] = \exp\{\mu + \sigma^2/2\}$.

Another family that only has closed expression for the density function is the renowned gamma family. This model has been widely used to model the working time of a unit in reliability.

Definition 1.2.9. Given a random variable X, it is said that X follows a *gamma distribution* with shape parameter $\alpha > 0$ and scale parameter $\beta > 0$, denoted by $X \sim G(\alpha, \beta)$, if its density function is given by

$$f(x) = \frac{x^{\alpha-1} \exp\left\{-\frac{x}{\beta}\right\}}{\beta^\alpha \Gamma(\alpha)}, \quad \text{for all } x \in (0, +\infty),$$

where Γ denotes the gamma function, and its mean is given by $E[X] = \alpha\beta$. The gamma distributions are ILR, whenever $\alpha \geq 1$ and, by (1.5), IFR as well. If $\alpha < 1$, then they are DFR.

This model has been generalized to a parametric family which contains multiple models as particular cases. This family has been mainly applied to fit income distributions [18, 19].

Definition 1.2.10. Given a random variable X, it is said that X follows a *generalized gamma distribution* with shape parameters $a, p > 0$ and scale parameter $\beta > 0$, denoted by $X \sim GG(a, \beta, p)$, if its density function is given by

$$f(x) = \alpha \frac{x^{ap-1} \exp\left\{-\left(\frac{x}{\beta}\right)^a\right\}}{\beta^{ap}\Gamma(p)}, \quad \text{for all } x \in (0, +\infty).$$

By letting $p = \alpha$ and $a = 1$, the gamma distribution arises as a particular case of this family.

There exists another family that generalizes several well-known models, the type II generalized beta distribution. Some of the parametric families that are generalized by this model are the Singh-Maddala, the Dagum, the Lomax, and the Fisk distributions (see Ref. [18] for more details).

Definition 1.2.11. Given a random variable X, it is said that X follows a *type II generalized beta distribution* with shape parameters $a, p, q > 0$ and scale parameter $b > 0$, denoted by $X \sim GB2(a, b, p, q)$, if its density function is given by

$$f(x) = \frac{a x^{ap-1}}{b^{ap} \beta(p, q) \left[1 + \left(\frac{x}{b}\right)^a\right]^{p+q}}, \quad \text{for all } x \in (0, +\infty),$$

where β denotes the beta function. Letting $q = 1$ and $p = 1$, the Dagum and the Singh-Maddala distributions arise as a particular cases of this family, respectively.

Next, several parametric families that also have explicit expressions for the survival and quantile functions are defined. Let us consider first the Pareto family, which is applied in reliability and risk theory (see, for instance, [20–22]).

Definition 1.2.12. Given a random variable X, it is said that X follows a *Pareto distribution* with shape parameter $a > 0$ and scale parameter $k \in \mathbb{R}$, denoted by $X \sim P(a, k)$, if its survival function is given by

$$\overline{F}(x) = \left(\frac{k}{x}\right)^a, \quad \text{for all } x \in (k, +\infty),$$

its quantile function is given by

$$F^{-1}(p) = \frac{k}{(1-p)^{\frac{1}{a}}}, \quad \text{for all } p \in (0, 1),$$

and its mean exists if $a > 1$ and is given by $E[X] = ak/(a-1)$. The Pareto distributions are DFR.

Another parametric family with these characteristics, that means, with explicit expression for both the survival and the quantile functions, is the Weibull distribution. This family is widely applied in reliability and risk theory [20, 21, 23].

Definition 1.2.13. Given a random variable X, it is said that X follows a *Weibull distribution* with shape parameter $\beta > 0$ and scale parameter $\alpha > 0$, denoted by $X \sim W(\alpha, \beta)$, if its survival function is given by

$$\overline{F}(x) = \exp\left\{-\left(\frac{x}{\alpha}\right)^{\beta}\right\}, \quad \text{for all } x \in (0, +\infty),$$

its quantile function is given by

$$F^{-1}(p) = \alpha(-\log(1-p))^{\frac{1}{\beta}}, \quad \text{for all } p \in (0, 1),$$

and its mean is given by $E[X] = \alpha\Gamma(1 + 1/\beta)$. The particular case for $\beta = 1$, it is known as the exponential family, denoted by $X \sim Exp(1/\alpha)$. The Weibull distributions are ILR, whenever $\beta \geq 1$ and, by (1.5), IFR as well. If $\beta < 1$, then they are DFR. Notice that letting $a = \beta$, $\beta = \alpha$ and $p = 1$, the Weibull distribution becomes a particular case of the generalized gamma distribution.

Below, two models are defined in terms of their quantile functions, which do not have closed expression for any other associated function. The first one, called the Davies distribution, is introduced by Hankin and Lee [24], and it has been applied to fit data in reliability and hydrology [25].

Definition 1.2.14. Given a random variable X, it is said that X follows a *Davies distribution* with shape parameters $\lambda, \theta > 0$ and scale parameter $C > 0$, denoted by $X \sim D(\lambda, \theta, C)$, if its quantile function is given by

$$F^{-1}(p) = C\frac{p^{\lambda}}{(1-p)^{\theta}}, \quad \text{for all } p \in (0, 1),$$

and its mean exists if $\theta < 1$ and is given by $E[X] = C\beta(1+\lambda, 1-\theta)$. Taking limits as λ tends to zero, the Pareto distribution arises as a limit case of this family, letting $C = k$ and $\theta = 1/a$.

The following parametric family, the Govindarajulu distribution, is introduced by Govindarajulu [26], and it has been already applied in reliability [27].

Definition 1.2.15. Given a random variable X, it is said that X follows a *Govindarajulu distribution* with shape parameter $\beta > 0$, scale parameter

$\sigma > 0$ and location parameter $\theta \geq 0$, denoted by $X \sim G(\beta, \sigma, \theta)$, if its quantile function is given by

$$F^{-1}(p) = \theta + \sigma \left((\beta + 1)p^{\beta} - \beta p^{\beta+1} \right), \quad \text{for all } p \in (0, 1),$$

and its mean is given by $E[X] = \theta + \sigma (1 - \beta(\beta + 2))$.

Next, some discrete parametric families are defined. First, we introduce the most well-known model, the binomial distribution.

Definition 1.2.16. Given a random variable X, it is said that X follows a *binomial distribution* with parameters $n > 0$ and $p \in (0, 1)$, denoted by $X \sim B(n, p)$, if its mass probability function is given by

$$p(x) = \binom{n}{x} p^x (1 - p)^{n-x}, \quad \text{for all } x = 0, 1, \dots, n.$$

The Bernoulli distribution arises as a particular case of this family, letting $n = 1$.

Definition 1.2.17. Given a random variable X, it is said that X follows a *negative binomial distribution* with parameters $r > 0$ and $p \in (0, 1)$, denoted by $X \sim BN(r, p)$, if its mass probability function is given by

$$p(x) = \binom{r + x - 1}{x} p^r (1 - p)^x, \quad \text{for all } x \in \{0\} \cup \mathbb{N}.$$

Definition 1.2.18. Given a random variable X, it is said that X follows a *Poisson distribution* with parameter $\lambda > 0$, denoted by $X \sim P(\lambda)$, if its mass probability function is given by

$$p(x) = \frac{\exp\{-\lambda\} \lambda^x}{x!}, \quad \text{for all } x \in \{0\} \cup \mathbb{N}.$$

Definition 1.2.19. Given a random variable X, it is said that X follows a *hypergeometric distribution* with parameters $N > 0$ and $n, m \leq N$, denoted by $X \sim H(N, m, n)$, if its mass probability function is given by

$$p(x) = \frac{\binom{m}{x}\binom{N-m}{n-x}}{\binom{N}{n}},$$

for all $n - \min\{n, N-m\} < x < \min\{n, m\}$.

1.3 MULTIVARIATE DISTRIBUTION NOTIONS

In this section, several notions and results for multivariate distributions are provided. Given a random vector $\mathbf{X} = (X_1, \ldots, X_n)$, it is defined its joint distribution function, denoted by F, as

$$F(\mathbf{x}) = P[\mathbf{X} \le \mathbf{x}] = P[X_1 \le x_1, \ldots, X_n \le x_n],$$

for all $\mathbf{x} = (x_1, \ldots, x_n) \in \mathbb{R}^n$. Analogously to the univariate case, if the *joint distribution function* can be represented as

$$F(\mathbf{x}) = \int_{\prod_{i=1}^{n}(-\infty, x_i]} f(\mathbf{u}) \, d\mathbf{u},$$

then $\mathbf{X}$ is said to be continuous with *joint density function f.* In this book, we are mainly interested in continuous random vectors, although most results are still true (or can be written) for discrete random vectors. The marginal distribution function of each component, denoted by F_i, can be derived from the joint distribution function as

$$F_i(x) = P[X_i \le x] = \lim_{\substack{x_j \to -\infty \\ j \ne i}} F(x_1, \ldots, x_{i-1}, x, x_{i+1}, \ldots, x_n).$$

However, the joint distribution function cannot be obtained, in general, from the marginal ones. In the particular case where the components are independent, we find that

$$F(\mathbf{x}) = \prod_{i=1}^{n} F_i(x_i).$$

As in the univariate case, the *joint survival function* is defined as

$$\overline{F}(x_1, \ldots, x_n) = P[X_1 > x_1, \ldots, X_n > x_n], \quad \text{for all } (x_1, \ldots, x_n) \in \mathbb{R}^n.$$

Recall that the usual componentwise partial order on $\mathbb{R}^n$ will be denoted by $\le$ and it means that, given $\mathbf{x} = (x_1, \ldots, x_n)$ and $\mathbf{y} = (y_1, \ldots, y_n)$, it holds $\mathbf{x} \le \mathbf{y}$, if $x_i \le y_i$, for all $i = 1, \ldots, n$. A function $f: A \subseteq \mathbb{R}^n \mapsto \mathbb{R}^k$ is said to be increasing [decreasing] if $f(\mathbf{x}) \le f(\mathbf{y})$, for all $\mathbf{x} \le [\ge]\mathbf{y}$.

1.3.1 The standard construction

The problem of defining a suitable multivariate extension of the univariate quantile function has a long history in statistics and probability. Next, the definition of the multivariate quantile transform is recalled. This transformation is also known as the standard construction, and is introduced by Arjas and Lehtonen [28], O'Brien [29], Rosenblatt [30], or Rüschendorf [31]. Given a continuous random vector $\mathbf{X} = (X_1, \ldots, X_n)$, its *standard construction* is recursively defined as

$$Q_{\mathbf{X},1}(p_1) = F_{X_1}^{-1}(p_1),$$

$$Q_{\mathbf{X},2}(p_1, p_2) = F_{[X_2|X_1=Q_{\mathbf{X},1}(p_1)]}^{-1}(p_2),$$

$$\vdots$$

$$Q_{\mathbf{X},n}(p_1, \ldots, p_n) = F_{\left[X_n \mid \bigcap_{j=1}^{n-1}\{X_j=Q_{\mathbf{X},j}(p_1,\ldots,p_j)\}\right]}^{-1}(p_n),$$

(1.11)

for all $(p_1, p_2, \ldots, p_n) \in (0,1)^n$, where $F_{X_1}^{-1}$ denotes the quantile function of X_1 and

$$F_{\left[X_i \mid \bigcap_{j=1}^{i-1}\{X_j=Q_{\mathbf{X},j}(p_1,\ldots,p_j)\}\right]}^{-1}, \qquad \text{for all } i = 2, \ldots, n,$$

denotes the quantile functions of the univariate conditional random variables given by

$$\left[X_i \,\middle|\, \bigcap_{j=1}^{i-1}\{X_j = Q_{\mathbf{X},j}(p_1, \ldots, p_j)\}\right], \qquad \text{for all } i = 2, \ldots, n.$$

It is worth mentioning that this well-known transform is widely used in simulation theory and it plays the role of the quantile in the multivariate case. Hence, given n independent uniformly distributed random variables $U_1, \ldots, U_n$ on the interval $(0, 1)$, then

$$(X_1, \ldots, X_n) =_{\text{st}} \mathbf{Q}_{\mathbf{X}}(U_1, \ldots, U_n),$$

(1.12)

where

$$\mathbf{Q}_{\mathbf{X}}(p_1, \ldots, p_n) = (Q_{\mathbf{X},1}(p_1), Q_{\mathbf{X},2}(p_1, p_2), \ldots, Q_{\mathbf{X},n}(p_1, \ldots, p_n)),$$

for all $(p_1, \ldots, p_n) \in (0,1)^n$. As we shall see later, another transform of interest is the multivariate distributional transform. Given a continuous random vector $\mathbf{X} = (X_1, \ldots, X_n)$, its *multivariate distributional transform* is recursively defined as

$$D_{\mathbf{X},1}(x_1) = F_{X_1}(x_1),$$

$$D_{\mathbf{X},2}(x_1,x_2) = F_{[X_2|X_1=x_1]}(x_2),$$

$$\vdots \tag{1.13}$$

$$D_{\mathbf{X},n}(x_1,\ldots,x_n) = F_{\left[X_n|\bigcap_{j=1}^{n-1}\{X_j=x_j\}\right]}(x_n),$$

for all $(x_1,\ldots,x_n)$ in the support of $\mathbf{X}$, where F_{X_1} denotes the distribution function of X_1 and

$$F_{\left[X_i|\bigcap_{j=1}^{i-1}\{X_j=x_j\}\right]}, \qquad \text{for all } i = 2,\ldots,n$$

denote the distribution function of the conditional random variables given by

$$\left[X_i\left|\bigcap_{j=1}^{i-1}\{X_j=x_j\}\right.\right], \qquad \text{for all } i = 2,\ldots,n.$$

According to the previous notation, we find that

$$(U_1,\ldots,U_n) =_{\text{st}} \mathbf{D_X}(X_1,\ldots,X_n), \tag{1.14}$$

where

$$\mathbf{D_X}(x_1,\ldots,x_n) = (D_{\mathbf{X},1}(x_1),D_{\mathbf{X},2}(x_1,x_2),\ldots,D_{\mathbf{X},n}(x_1,\ldots,x_n)),$$

for all $(x_1,\ldots,x_n) \in \mathbb{R}^n$. Given two continuous random vectors $\mathbf{X} = (X_1,\ldots,X_n)$ and $\mathbf{Y} = (Y_1,\ldots,Y_n)$, we can consider the transformation

$$\Phi(x_1,\ldots,x_n) = \mathbf{Q_Y}(\mathbf{D_X}(x_1,\ldots,x_n)), \tag{1.15}$$

defined for all $(x_1,\ldots,x_n)$ in the support of $\mathbf{X}$. Observe that

$$\Phi_1(x_1) = F_{Y_1}^{-1}(F_{X_1}(x_1)),$$

$$\Phi_2(x_1,x_2) = F_{[Y_2|Y_1=\Phi_1(x_1)]}^{-1}\left(F_{[X_2|X_1=x_1]}(x_2)\right),$$

$$\vdots$$

$$\Phi_n(x_1,\ldots,x_n) = F_{\left[Y_n|\bigcap_{j=1}^{n-1}\{Y_j=\Phi_j(x_1,\ldots,x_j)\}\right]}^{-1}\left(F_{\left[X_n|\bigcap_{j=1}^{n-1}\{X_j=x_j\}\right]}(x_n)\right),$$

for all $(x_1,\ldots,x_n)$ in the support of $\mathbf{X}$. Since the distribution and quantile functions are increasing, then $\Phi_i(x_1,\ldots,x_i)$ is also increasing in x_i, for all $i = 1,\ldots,n$. In fact, in cases of differentiability, the Jacobian matrix of Φ is always a lower triangular matrix with strictly positive diagonal elements.

The most relevant property of Φ is that, from (1.12) and (1.14), we have

$$\mathbf{Y} =_{\mathrm{st}} \Phi(\mathbf{X}),$$

and, consequently, the function Φ maps the random vector $\mathbf{X}$ onto $\mathbf{Y}$.

In addition, Fernández-Ponce and Suárez-Llorens [32] prove in their Theorem 3.1 that if we take a function $\mathbf{k} : \mathbb{R}^n \mapsto \mathbb{R}^n$ such that $\mathbf{Y} =_{\mathrm{st}} \mathbf{k}(\mathbf{X})$ and $\mathbf{k}$ has a lower triangular Jacobian matrix with strictly positive diagonal elements, then $\mathbf{k}$ necessarily has the form of the function Φ given in (1.15).

1.3.2 Copulas

As has been mentioned previously, the marginal distributions do not provide enough information, in general, to obtain the joint distribution function. An additional element is required: the copula.

The notion of copulas is introduced by Sklar [33] and is studied by Kimeldorf and Sampson [34], among others, under the name of uniform representation, and by Deheuvels [35] under the name of dependence function. The term copula was first used by Sklar [33] and is derived from the Latin word *copulare*, which means to connect or to join. The main purpose of copulas is to describe the dependence structure of a random vector.

Copulas have become a popular multivariate modeling tool in many fields where multivariate dependence has interest, since these functions describe in a more complete way the dependence structure among the components of a random vector.

In actuarial science, copulas are used to model dependent mortality and losses [36–38]. In finance, copulas are used in asset allocation, credit scoring, derivative pricing, and risk management [39–41]. In biomedical studies, copulas are used to model correlated event times and competing risks [42, 43]. In engineering, copulas are used in multivariate process control and hydrological modeling [44, 45].

Let us give the formal definition. A *copula* $C : [0, 1]^n \mapsto [0, 1]$ is a cumulative distribution function with uniformly distributed marginal functions on $[0, 1]$.

Sklar [33] gives one of the main results on copulas. On the one hand, there always exists a copula associated to any multivariate distribution function. On the other hand, any copula, evaluated in certain marginal distribution functions, leads to a multivariate distribution function with such marginals. In particular, Sklar [33] states the following result.

Theorem 1.3.1. *Let C be a copula and F_i be n distribution functions. Then, the function*

$$F(x_1, \ldots, x_n) = C(F_1(x_1), \ldots, F_n(x_n)), \quad \text{for all } (x_1, \ldots, x_n) \in \mathbb{R}^n,$$

is a multivariate distribution function with marginals $F_1, \ldots, F_n$. Furthermore, if F is the joint distribution function of a random vector with continuous marginals $F_1, \ldots, F_n$, then C is unique and is given by

$$C(p_1, \ldots, p_n) = F(F_1^{-1}(p_1), \ldots, F_n^{-1}(p_n)), \quad \text{for all } (p_1, \ldots, p_n) \in [0, 1]^n.$$

There is a great variety of parametric families of copulas, and we refer the reader to Ref. [46] for examples and more details about copulas.

1.3.3 The multivariate dynamic hazard rate and mean residual life functions

Next, multivariate extensions of the hazard rate and mean residual life functions are introduced. In the literature, several definitions can be found for these functions. The ones that we are going to consider are the multivariate dynamic versions introduced by Shaked and Shanthikumar [47, 48]. For a recent review, the reader can look at Shaked and Shanthikumar [49]. These functions require the "history" notion, which will be recalled next.

Let us consider a random vector $\mathbf{X} = (X_1, \ldots, X_n)$ where X_i represents the lifetimes of n units, therefore the components are assumed to be non-negative. For $t \geq 0$, let h_t denote the list of units which have failed and their failure times, which is called a *history*. Down to the last detail,

$$h_t = \{\mathbf{X}_I = \mathbf{x}_I, \mathbf{X}_{\bar{I}} > t\mathbf{e}\},$$

where $I = \{i_1, \ldots, i_k\} \subseteq \{1, \ldots, n\}, \bar{I} = \{1, \ldots, n\} \setminus I, \mathbf{X}_I$ denotes the vector formed by the components of $\mathbf{X}$ with index in I and $0 < x_{i_j} < t$, for all $j = 1, \ldots, k$, and $\mathbf{e}$ denotes a vector of 1, where the dimension is determined from the context. In this case, the dimension of $\mathbf{e}$ is equal to $n - k$.

Now, we proceed to give the definition of the multivariate hazard rate function. Given a history h_t as above and $j \in \bar{I}$, the *multivariate dynamic hazard rate function* of X_j given the history h_t, is defined by

$$\eta_j(t|h_t) = \lim_{\Delta \to 0^+} \frac{1}{\Delta} P[t < X_j \le t + \Delta | h_t], \quad \text{for all } t \ge 0. \tag{1.16}$$

Clearly, $\eta_j(t|h_t)$ is the intensity of failure of the component j, given the history h_t, or the failure rate of X_j at time t given h_t.

Finally, we recall the definition of the multivariate mean residual life function. Given a random vector $\mathbf{X}$, a history $h_t = \{\mathbf{X}_I = \mathbf{x}_I, \mathbf{X}_{\bar{I}} > t\mathbf{e}\}$, and $j \in \bar{I}$, the *multivariate dynamic mean residual function* of X_j given h_t, is defined by

$$m_j(t|h_t) = E[X_j - t|h_t], \quad \text{for all } t \ge 0.$$

In this case, $m_j(t|h_t)$ is the mean residual life function of X_j at time t given h_t.

1.3.4 Dependence notions

Next, some definitions of positive dependence are recalled (see, e.g., Ref. [50] for a complete revision of these notions). These notions will be used as a tool for stating some results in Chapter 3.

Given a random vector $\mathbf{X} = (X_1, \dots, X_n)$, we say that $\mathbf{X}$ is *conditionally increasing in sequence*, denoted by CIS, if the distribution function of the random variable

$$[X_i | X_1 = x_1, \dots, X_{i-1} = x_{i-1}]$$

is decreasing in $(x_1, \dots, x_{i-1})$ in the support of $(X_1, \dots, X_{i-1})$, for all $i = 2, \dots, n$. Moreover, it is said that a random vector $\mathbf{X} = (X_1, \dots, X_n)$ is *conditionally increasing*, denoted by CI, if the random vector $\mathbf{X}_\pi = (X_{\pi(1)}, \dots, X_{\pi(n)})$ is CIS, for all permutations π. Notice that CI is stronger than CIS.

Let us see a weaker positive dependence property. Given a random vector $\mathbf{X} = (X_1, \dots, X_n)$, we say that $\mathbf{X}$ is *positive associated*, denoted by PA, if

$$E[h_1(\mathbf{X})h_2(\mathbf{X})] \ge E[h_1(\mathbf{X})]E[h_2(\mathbf{X})],$$

for all increasing functions $h_i : \mathbb{R}^n \mapsto \mathbb{R}$, for all $i = 1, 2$. Since the PA property is preserved under increasing transformations (see, for instance, Ref. [51]), then, given a PA copula, any vector with copula C is also PA. The same result holds for the CI property.

Finally, the notion of multivariate total positivity of order 2 is recalled. Given a random vector $\mathbf{X} = (X_1, \dots, X_n)$ with joint density function f, it is said that $\mathbf{X}$ is *multivariate totally positive of order 2*, denoted by MTP_2, if

$$f(\mathbf{x} \vee \mathbf{y})f(\mathbf{x} \wedge \mathbf{y}) \geq f(\mathbf{x})f(\mathbf{y}), \quad \text{for all } \mathbf{x}, \mathbf{y} \in \mathbb{R}^n,$$

where

$$\mathbf{x} \vee \mathbf{y} = (x_1 \vee y_1, \ldots, x_n \vee y_n)$$

and

$$\mathbf{x} \wedge \mathbf{y} = (x_1 \wedge y_1, \ldots, x_n \wedge y_n),$$

where $\vee$ and $\wedge$ denote the minimum and the maximum operators.

One of the main uses of this property is as a tool to check the CIS, CI, and PA properties, which is difficult in many situations. In particular, the following implications are well known:

$$\text{MTP}_2 \Rightarrow \text{CI} \Rightarrow \text{CIS} \Rightarrow \text{PA}.$$

1.3.5 Parametric families of multivariate distributions

In this section, some known multivariate distributions are recalled. These models will be compared in Chapter 3 in some multivariate stochastic orders.

It is worth mentioning that the mean of a random vector $\mathbf{X} = (X_1, \ldots, X_n)$ is given by

$$E[\mathbf{X}] = \begin{pmatrix} E[X_1] \\ \vdots \\ E[X_n] \end{pmatrix},$$

and the covariance matrix of $\mathbf{X}$ is

$$\text{Cov}(\mathbf{X}) = E[(\mathbf{X} - E[\mathbf{X}])(\mathbf{X} - E[\mathbf{X}])^{\mathsf{T}}].$$

Next, we consider some multivariate distributions. First, the definition of the multivariate normal distribution is recalled.

Definition 1.3.2. Given a random vector $\mathbf{X} = (X_1, \ldots, X_n)$, it is said that $\mathbf{X}$ follows a *multivariate normal distribution* with mean vector $\mu \in \mathbb{R}^n$ and covariance matrix $\Sigma \in \mathbb{R}^n \times \mathbb{R}^n$, denoted by $\mathbf{X} \sim N_n(\mu, \Sigma)$, if its joint density function is given by

$$f(\mathbf{x}) = \frac{1}{(2\pi)^{\frac{n}{2}}|\Sigma|^{\frac{1}{2}}} \exp\left\{-\frac{1}{2}(\mathbf{x}-\mu)^{\mathrm{T}}\Sigma^{-1}(\mathbf{x}-\mu)\right\}, \quad \text{for all } \mathbf{x} \in \mathbb{R}^{n}.$$

The marginal distribution functions follow univariate normal models. Furthermore, the copula of a $N_n(\mu, \Sigma)$ is the same to that of $N_n(\mathbf{0}, \mathbf{P})$ where $\mathbf{P}$ is the correlation matrix obtained through the covariance matrix Σ. In this sense, all multivariate normal distributions with the same dimension and correlation matrix have the same (Gaussian) copula.

Another well-known multivariate family is the elliptically contoured model. Let us consider the formal definition.

Definition 1.3.3. Given a random vector $\mathbf{X} = (X_1, \ldots, X_n)$, it is said that $\mathbf{X}$ follows an *elliptically contoured distribution*, denoted by $E_n(\mu, \Sigma, g)$, if its joint density function is given by

$$f(\mathbf{x}) = \frac{1}{\sqrt{|\Sigma|}} g((\mathbf{x}-\mu)'\Sigma^{-1}(\mathbf{x}-\mu)), \quad \text{for all } \mathbf{x} \in \mathbb{R}^{n}, \tag{1.17}$$

where μ is the median vector (which is also the mean vector if the latter exists), Σ is a symmetric positive definite matrix which is proportional to the covariance matrix, if the latter exists, and $g : \mathbb{R}_+ \mapsto \mathbb{R}_+$ such that $\int g(x)\,dx < +\infty$.

A particular case of elliptically contoured distributions is the case of a multivariate normal taking

$$g(x) = \frac{1}{(2\pi)^n} \exp\left\{-\frac{1}{2}x\right\}.$$

Some other particular cases are described in Ref. [52].

1.4 SUMMARY

This chapter gives an overview of some preliminary notions and results for univariate and multivariate distributions discussed in subsequent chapters. In Section 1.2, we introduced several functions and measures that are commonly used to define and characterize univariate stochastic orders. The fields where these functions and measures play a relevant role have been also highlighted, like reliability, survival analysis, risks, and economics. As a theoretical tool for some proofs in this book, some results on total positivity theory have been provided. Section 1.2 finished with a review on some of the most prominent parametric families of univariate distributions, since we shall make a review on results for the stochastic comparison within these parametric families. In Section 1.3, we focused the attention on multivariate distributions. In this case we gave several functions related to the dependence structure of multivariate distributions, such as the standard construction and the copula notions. The multivariate dynamic hazard rate and mean residual life functions were introduced as well. As a tool for several results, some dependence notions were also introduced. Finally, several parametric families of multivariate distributions were given.

CHAPTER 2

Univariate stochastic orders

2.1 INTRODUCTION

One of the main objectives of statistics is the comparison of random quantities. These comparisons are mainly based on the comparison of some measures associated to these random quantities. For example, it is very common to compare two random variables in terms of their means, medians or variances. In some situations, comparisons based only on two single measures are not very informative. For instance, let us consider two Weibull distributed random variables X and Y with distribution functions $F(x) = 1 - \exp\left\{-x^2\right\}$ and $G(x) = 1 - \exp\left\{-\frac{2}{\sqrt{\pi}}x\right\}$, for all $x \geq 0$, respectively. In this case, we see that $E[X] = E[Y] = \sqrt{\pi}/2$. If X and Y represent the random lifetimes of two devices, or the survival lifetimes of patients under two different treatments, then we would say that X has the same expected survival time than Y, if we just considered the mean values. However, if we took into account the probability of surviving at a fixed time $x \geq 0$, then $P[X > x] \geq P[Y > x]$, for all $x \in [0, 2/\sqrt{\pi}]$ and $P[X > x] \leq P[Y > x]$, for all $t \in [2/\sqrt{\pi}, +\infty)$ (see Figure 2.1).

Therefore, the information provided by the survival functions is more complete than those provided by the means. The necessity of providing more detailed comparisons of two random quantities has motivated the development of the theory of stochastic orders, which has grown significantly during the last 50 years. The purpose of this chapter is to provide the reader with an introduction to some of the most popular comparison criteria including some properties of interest, fields where they can be applied, examples and some preliminary tools for the empirical validation of such criteria. Along the chapter, these criteria are named, in general, stochastic orders and they are based on the comparison of the functions defined in Section 1.2. The general organization in these sections is the following. First of all, the definitions and some characterizations are given. Secondly, some sets of sufficient conditions are provided (which are also applied to order several parametric families), as well as some preservation results. Finally, some graphical procedures are given to compare two data sets. To conclude this chapter, several applications of the stochastic orders are given in Section 2.9.

An Introduction to Stochastic Orders. http://dx.doi.org/10.1016/B978-0-12-803768-3.00002-8

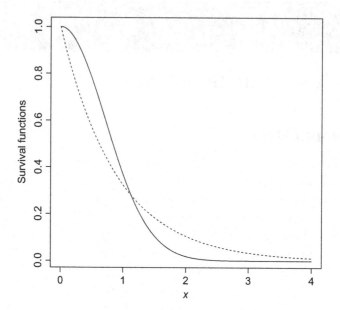

Figure 2.1 Survival functions of $X \sim W(1,2)$ (continuous line) and $Y \sim W(\sqrt{\pi}/2, 1)$ (dashed line).

In particular, several tables are included, which collect the sufficient conditions on the parameters of some family distributions so that the stochastic orders hold among two distributions belonging to the same model. Besides, some applications in reliability and risk theory are given. Unless stated otherwise, the main references for the results provided in this chapter are Refs. [2, 3].

2.2 THE USUAL STOCHASTIC ORDER

As was mentioned in the introduction, the theory of stochastic orders arises because of the fact that comparisons based only on single measures are not very informative. As seen in Section 1.2, if the random variable X represents the random lifetime of a device, or an organism, the survival function $\overline{F}(x)$ is a function of interest in this context. If we have another random lifetime Y with survival function $\overline{G}(x)$, then it is of interest to study if one of the two survival functions lies above the other one. This basic idea is the motivation to define the usual stochastic order. The formal definition is as follows.

Definition 2.2.1. Given two random variables X and Y, we say that X is smaller than Y in the *stochastic order*, denoted by $X \leq_{st} Y$, if

$$\overline{F}(x) \leq \overline{G}(x), \quad \text{for all } x \in \mathbb{R}.$$

We shall refer to this criterion as the stochastic order or st order indistinctly, taking into account the notation $\leq_{st}$, and analogously to the remainder criteria.

The definition of the stochastic order formalizes the idea that the random variable X is less likely than Y to take on large values and, therefore, compares the location of the random variables. The simplicity of the definition of this criterion allow us to check it directly, whenever the survival functions have explicit expressions. Let us compare two Pareto distributions.

Example 2.2.2. Let $X \sim P(a_1, k_1)$ and $Y \sim P(a_2, k_2)$ with survival functions $\overline{F}$ and $\overline{G}$, respectively. If we assume $k_2 \geq k_1$, it is easy to see that $\overline{F}(x) \leq \overline{G}(x)$, for all $x \leq k_2$. Additionally, if $a_2 \leq a_1$, there is no crossing point among $\overline{F}$ and $\overline{G}$, for all $x \geq k_2$. To sum up, if $k_2 \geq k_1$ and $a_2 \leq a_1$, then $X \leq_{st} Y$. Figure 2.2 shows a particular example of this situation.

Obviously, two survival functions can cross each other as it occurs in the example provided in the introduction and, consequently, the stochastic

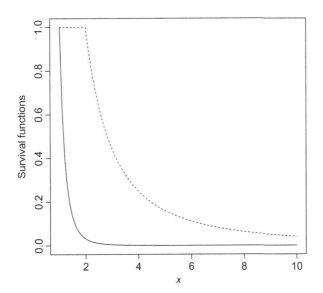

Figure 2.2 Survival functions of $X \sim P(5, 1)$ (continuous line) and $Y \sim P(2, 2)$ (dashed line).

order is a partial criterion on the set of distribution functions. In fact, it holds that:

(i) $X \leq_{st} X$ (the stochastic order is *reflexive*),

(ii) if $X \leq_{st} Y$ and $Y \leq_{st} Z$, then $X \leq_{st} Z$ (the stochastic order is *transitive*), and

(iii) $X \leq_{st} Y$ and $X \geq_{st} Y$ do not hold unless $X =_{st} Y$ (the stochastic order is *antisymmetric*).

All the stochastic orders considered in this chapter satisfy the properties (i) and (ii), but some of them fail to satisfy the antisymmetric property.

It is interesting to notice that the stochastic order is also related to the comparison of the VaR measure (see Section 1.2). In particular, the following theorem is established.

Theorem 2.2.3. *Let X and Y be two random variables with quantile functions F^{-1} and G^{-1}, respectively. Then, $X \leq_{st} Y$ if, and only if,*

$$F^{-1}(p) \leq G^{-1}(p), \quad \text{for all } p \in (0, 1).$$

Proof. Let us assume first that $X \leq_{st} Y$ and let $p \in (0, 1)$. Recalling the property (i) of the quantile function and from the inequality $F(x) \geq G(x)$, we see that

$$[F^{-1}(p), +\infty) = \{x \in \mathbb{R} | F(x) \geq p\}$$

$$\supseteq \{x \in \mathbb{R} | G(x) \geq p\} = [G^{-1}(p), +\infty),$$

therefore, $F^{-1}(p) \leq G^{-1}(p)$, for all $p \in (0, 1)$.

Let us assume now that $F^{-1}(p) \leq G^{-1}(p)$, for all $p \in (0, 1)$. Let us consider a uniformly distributed random variable U on the interval $(0, 1)$. From the assumption, we see that $\{F^{-1}(U) \leq x\} \supseteq \{G^{-1}(U) \leq x\}$, for all $x \in \mathbb{R}$. Therefore, taking into account the property (ii) of the quantile function, we conclude that

$$F(x) = P[F^{-1}(U) \leq x] \geq P[G^{-1}(U) \leq x] = G(x), \quad \text{for all } x \in \mathbb{R}.$$

$\square$

Recalling the notation for the value-at-risk notion, we see that $X \leq_{st} Y$ if, and only if, VaR$[X;p] \leq$ VaR$[Y;p]$, for all $p \in (0, 1)$. Therefore, the stochastic order is not only interesting in the context of reliability, but also in the context of risk theory. Another characterization that will be used is the following.

Theorem 2.2.4. *Let X and Y be two random variables with quantile functions F^{-1} and G^{-1}, respectively. Then $X \leq_{st} Y$ if, and only if,*

$$\min\{X, F^{-1}(p)\} \leq_{st} \min\{Y, G^{-1}(p)\}, \quad \textit{for all } p \in (0, 1).$$

Proof. The proof follows from the previous theorem and observing that the quantile function of $\min\{X, F^{-1}(p)\}$ is given by

$$F_p^{-1}(q) = \begin{cases} F^{-1}(q) & \text{if } 0 < q < p, \\ \\ F^{-1}(p) & \text{if } p \leq q < 1, \end{cases} \tag{2.1}$$

and analogously for $\min\{Y, G^{-1}(p)\}$.   □

From the interpretation of the random variable $\min\{X, F^{-1}(p)\}$ in risk theory and reliability, this characterization provides further applications of the stochastic order in these contexts. In finance and economics, the stochastic order is known as *first stochastic dominance* and it has been widely used. For specific applications in these contexts, we refer the reader to Refs. [7, 8].

Some other useful characterizations of the stochastic order are provided next.

Theorem 2.2.5. *Let X and Y be two random variables. Then $X \leq_{st} Y$ if, and only if, one of the following equivalent conditions holds:*

(i) Two random variables $\hat{X}$ and $\hat{Y}$ exist, defined on the same probability space, such that $\hat{X} =_{st} X$, $\hat{Y} =_{st} Y$ and $P[\hat{X} \leq \hat{Y}] = 1$.
(ii) $\phi(X) \leq_{st} \phi(Y)$, for all real valued increasing function ϕ.
(iii) $E[\phi(X)] \leq E[\phi(Y)]$, for all real valued increasing function ϕ such that the previous expectations exist.

Proof. First, we prove the equivalence among $X \leq_{st} Y$ and condition (i). Let us assume $X \leq_{st} Y$ and consider a random variable U uniformly distributed on the interval $(0, 1)$. Let F and G denote the distribution functions of X and Y, respectively. From the property (ii) of the quantile function, we see that $F^{-1}(U) =_{st} X$ and $G^{-1}(U) =_{st} Y$, and by Theorem 2.2.3, we find that $P[F^{-1}(U) \leq G^{-1}(U)] = 1$. Next, we prove the inverse implication. Let us assume now that two random variables $\hat{X}$ and $\hat{Y}$ exist defined on the same probability space, such that $\hat{X} =_{st} X$, $\hat{Y} =_{st} Y$ and $P[\hat{X} \leq \hat{Y}] = 1$. Since $\overline{F}(x) = P[\hat{X} > x]$ and $\overline{G}(x) = P[\hat{Y} > x]$, and $\{w \in \Omega | \hat{X}(w) > x\} \subseteq \{w \in \Omega | \hat{Y}(w) > x\}$, for all $x \in \mathbb{R}$, then $\overline{F}(x) = P[\hat{X} > x] \leq \overline{G}(x) = P[\hat{Y} > x]$, for all $x \in \mathbb{R}$.

Now, let us prove the equivalence with condition (ii). Let us assume $X \leq_{st} Y$ and consider the random variables $\hat{X}$ and $\hat{Y}$, described above, and any real valued increasing function ϕ. Since $\phi(\hat{X}) =_{st} \phi(X)$, $\phi(\hat{Y}) =_{st} \phi(Y)$ and $P[\phi(\hat{X}) \leq \phi(\hat{Y})] = 1$, then $\phi(X) \leq_{st} \phi(Y)$. The inverse implication follows by taking $\phi(x) = x$.

Finally, we prove the equivalence with condition (iii). Let us assume $X \leq_{st} Y$. From the equivalence provided by condition a) and the monotonicity of the expectation, we see that $E[X] = E[\hat{X}] \leq E[\hat{Y}] = E[Y]$. Now, a combination of this fact with the characterization provided by condition (ii) shows that $E[\phi(X)] \leq E[\phi(Y)]$, for all real valued increasing function ϕ such that $E[\phi(X)]$ and $E[\phi(Y)]$ exist. The inverse implication follows observing that $\overline{F}(x) = E[I_{(x,+\infty)}(X)]$ and $\overline{G}(x) = E[I_{(x,+\infty)}(Y)]$, for all $x \in \mathbb{R}$ and the fact that $I_{(x,+\infty)}$ is an increasing function.   □

These characterizations are of interest from both a theoretical and an applied point of view. The characterization provided by Theorem 2.2.5(i) is useful to prove results for the stochastic order. Let us observe that this characterization is closely related to the simulation of random variables. The characterization provided by Theorem 2.2.5(ii) shows that the stochastic order is preserved under increasing transformations, which is useful if it is easier to provide a comparison of X and Y rather than increasing transformations of the random variables, or one may be just more interested in some transformations of the random variables. For example, in such cases where $\phi(X)$ represents the benefit of a machine which depends increasingly on the random lifetime of the machine, represented by X. The condition provided by Theorem 2.2.5(iii) also highlights a way to compare random variables. The idea is to consider the comparison of two random variables in terms of expectations of transformations of the two random variables when such transformations belong to some specific family of functions. For a general theory of this approach, we refer the reader to Ref. [53]. It is also worth mentioning that a direct consequence of Theorem 2.2.5(iii) is the following

$$X \leq_{st} Y \Rightarrow E[X] \leq E[Y]. \tag{2.2}$$

Next, a sufficient condition for the stochastic order in terms of the density functions is provided, which is particularly useful in such situations where neither the quantile nor the survival functions have closed expressions.

Theorem 2.2.6. *Let X and Y be two continuous random variables with density functions f and g, respectively. If $S^-(g - f) = 1$ with the sign sequence $-, +$, then*

$$X \leq_{st} Y.$$

Proof. The proof follows directly applying Corollary 1.2.4 and from the sign sequence $-, +$ and the fact that the difference of the survival functions finishes with the same sign as the difference of the density functions. $\square$

The previous theorem is still valid for discrete random variables, replacing the density function by the mass probability function.

The stochastic order is preserved under some other operations like convergence, mixtures and convolution. First, a result on preservation under convergence is provided.

Theorem 2.2.7. *Let $\{X_n\}_{n \in \mathbb{N}}$ and $\{Y_n\}_{n \in \mathbb{N}}$ be two sequences of random variables such that X_n converges in distribution to X and Y_n converges in distribution to Y. If $X_n \leq_{st} Y_n$, for all $n \in \mathbb{N}$, then*

$$X \leq_{st} Y.$$

Proof. Let us denote by F and G the distribution functions of X and Y, respectively. If x is a point of continuity of F and G, we see that $\overline{F}(x) \leq \overline{G}(x)$, by taking limits and from the assumption $X_n \leq_{st} Y_n$. Since the set of points of continuity of F and G is dense in $\mathbb{R}$, and the distribution functions are right continuous, we conclude that $\overline{F}(x) \leq \overline{G}(x)$, for all $x \in \mathbb{R}$. $\square$

The following result deals with the preservation of the stochastic order under mixtures.

Theorem 2.2.8. *Let $\{X_1(\theta) | \theta \in S \subseteq \mathbb{R}\}$ and $\{X_2(\theta) | \theta \in S \subseteq \mathbb{R}\}$ be two families of random variables, and Θ_1 and Θ_2 be two random variables with common support S. If*

$$X_1(\theta) \leq_{st} X_2(\theta), \text{ for all } \theta \in S, \tag{2.3}$$

$$\Theta_1 \leq_{st} \Theta_2, \tag{2.4}$$

and $E[\phi(X_1(\theta))]$ or $E[\phi(X_2(\theta))]$ or both are increasing in θ, for all real valued increasing function ϕ, then

$$X_1(\Theta_1) \leq_{st} X_2(\Theta_2).$$

Proof. Let us consider an increasing real valued function ϕ such that $E[\phi(X_1(\Theta_1))]$ and $E[\phi(X_2(\Theta_2))]$ exist. First, we recall that $E[\phi(X_1(\Theta_1))] = E[l_1(\Theta_1)]$ and $E[\phi(X_1(\Theta_1))] = E[l_2(\Theta_2)]$, where $l_1(\theta) = E[\phi(X_1(\theta))]$ and $l_2(\theta) = E[\phi(X_2(\theta))]$. Let us consider that $E[l_1(\theta)]$ is increasing in θ, the case for $E[l_2(\theta)]$ is similar. Then, we see that

$$E[\phi(X_1(\Theta_1))] = E[l_1(\Theta_1)] \leq E[l_1(\Theta_2)] \leq E[l_2(\Theta_2)] = E[\phi(X_2(\Theta_2))],$$

where the first inequality follows combining (2.4) with the fact that $E[f_1(\theta)]$ is increasing in θ, and the second one combining (2.3) with the monotonicity of the expectation. $\qquad\square$

Finally, the preservation result under convolutions is considered. A more general result will follow from the properties of the multivariate stochastic order (see Section 3.2).

Theorem 2.2.9. *Let $X_1, \ldots, X_n$ and $Y_1, \ldots, Y_n$ be two sets of independent random variables. If $X_i \leq_{st} Y_i$, for all $i = 1, \ldots, n$, then*

$$\sum_{i=1}^{n} X_i \leq_{st} \sum_{i=1}^{n} Y_i.$$

Proof. The proof follows by induction, taking into account the following argument. Let us consider X_1, X_2 and Y_1, Y_2. Then, clearly

$$[X_1 + X_2 | X_2 = x] \leq_{st} [Y_1 + Y_2 | Y_2 = x], \quad \text{for all } x \in \mathbb{R},$$

and $E[\phi(X_1 + X_2 | X_2 = x)]$ and $E[\phi(Y_1 + Y_2 | Y_2 = x)]$ are increasing in x, for all ϕ increasing. Therefore, from $X_2 \leq_{st} Y_2$, the result follows from the previous theorem. $\qquad\square$

One of the purposes of this book is to illustrate, with some real data sets, examples where the different stochastic orders hold. Since this is an introductory book, we only consider preliminary techniques for the validation of the stochastic orders. In this case, we only consider non-parametric estimations of the different functions involved in the comparison. Additionally, we can consider hypothesis testing techniques; however, this would require another book on the topic, and so it will be not considered here.

In particular, given two independent samples from two random variables, a first step for the empirical validation of the stochastic order among the parent populations, is to plot the empirical survival functions. Let us consider the following example for two data sets taken from Hoel [54]. The data set comes from two groups of survival times of RFM strain male mice. Hoel [54] considers three main groups depending on the cause of death. We consider here the group where the cause of death is different from thymic lynfoma and cell sarcoma, which is labeled "other causes." This group is divided in two subgroups. The first subgroup lives in a conventional laboratory environment, which will be denoted by OCLE, while the second subgroup lives in a germ free environment, which will be denoted by OGFE. The reported data are the time to death in days. Figure 2.3 provides empirical evidence for the stochastic order among these two subgroups.

Clearly, another alternative arises from the plot of the curve $(\overline{F}(x), \overline{G}(x))$, that is, the $\overline{P} - \overline{P}$ plot. If $X \leq_{st} Y$, then the points of the $\overline{P} - \overline{P}$ plot should lie above the diagonal $x = y$. Notice that, if X and Y are continuous, the parametric curve described by the $\overline{P} - \overline{P}$ plot is the same than the plot of the real valued function $\overline{G}(\overline{F}^{-1}(t))$.

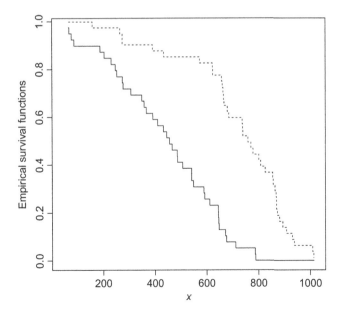

Figure 2.3 Empirical survival functions for OCLE (continuous line) and OGFE (dashed line).

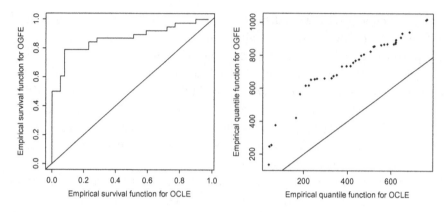

Figure 2.4 Empirical $\overline{P} - \overline{P}$ plot on the left side and empirical $Q - Q$ plot on the right side, for OCLE and OGFE.

Analogously, we can also check the stochastic order by plotting the quantiles functions, $(F^{-1}(p), G^{-1}(p))$, that is, the $Q - Q$ plot. We see that $X \leq_{st} Y$, if the points of the $Q - Q$ plot lie above the diagonal $x = y$.

Given two samples, we can consider the empirical $\overline{P} - \overline{P}$ and $Q - Q$ plots provided through the empirical distribution and quantile functions of the two samples. If we consider the example provided previously for the two groups of survival times of RFM strain male mice, the empirical $\overline{P} - \overline{P}$ and $Q - Q$ plots suggest the stochastic order among the two populations—see Figure 2.4.

2.3 THE INCREASING CONVEX ORDER AND RELATED ORDERS

As was pointed out in the previous section, the stochastic order does not always hold and, therefore, it is a partial order. Consequently, it is important to provide more tools to compare random variables. According to Theorem 2.2.5(iii), the stochastic order is characterized by the comparison of the expectations of increasing transformations of the random variables. If we restrict our attention to a subset of these transformations (that are still of interest), then it is possible to provide a weaker partial criterion to compare random variables, which is the case of the increasing convex order. Let us see its formal definition.

Definition 2.3.1. Given two random variables X and Y, we say that X is smaller than Y in the *increasing convex order*, denoted by $X \leq_{icx} Y$, if

$$E[\phi(X)] \leq E[\phi(Y)],$$

for all real valued increasing convex function ϕ such that the previous expectations exist.

This partial order can be also stated in terms of the stop-loss function (see Section 1.2). Since $(t - x)_+$ is an increasing and convex function, for any fixed value $x \in \mathbb{R}$, we see that $X \leq_{\text{icx}} Y$ implies

$$E[(X - x)_+] \leq E[(Y - x)_+], \text{ for all } x \in \mathbb{R}.$$

The inverse is also true, as we shall see in the following theorem. Since the existence of the stop-loss function requires the random variable to have finite mean, this condition will be assumed in this section.

Theorem 2.3.2. *Let X and Y be two random variables with finite means. Then, $X \leq_{\text{icx}} Y$ if, and only if,*

$$E[(X - x)_+] \leq E[(Y - x)_+], \quad \text{for all } x \in \mathbb{R}.$$

Proof. The proof follows observing that every increasing convex function can be obtained as the limit of positive linear combinations of functions $\phi_d(t) = (t - d)_+$. □

Notice that, from the previous theorem and (1.1), we see that $X \leq_{\text{icx}} Y$ if, and only if,

$$\int_x^{+\infty} \overline{F}(u)\, du \leq \int_x^{+\infty} \overline{G}(u)\, du, \quad \text{for all } x \in \mathbb{R}. \tag{2.5}$$

From the previous result, one of the main applications of the icx order is the comparison of the stop-loss functions in risk theory. As we shall see later, the icx order is also related to the comparison of some other risk measures.

Next, a characterization of the increasing convex order by integrals of the quantiles functions is provided. Its proof relies on some arguments and notions of measure theory and it is omitted. For more details, see Ref. [55].

Theorem 2.3.3. *Let X and Y be two random variables with quantile functions F^{-1} and G^{-1}, respectively, and finite means. Then, $X \leq_{\text{icx}} Y$ if, and only if,*

$$\int_p^1 F^{-1}(u)\, du \leq \int_p^1 G^{-1}(u)\, du, \quad \text{for all } p \in (0, 1).$$

From the definition of the tail value at risk and the conditional tail expectation (see Section 1.2), we see that $X \leq_{\text{icx}} Y$ if, and only if,

$$\text{TVaR}[X; p] \leq \text{TVaR}[Y; p], \quad \text{for all } p \in (0, 1),$$

which is equivalent to

$$\text{CTE}[X; p] \leq \text{CTE}[Y; p], \quad \text{for all } p \in (0, 1),$$

in the continuous case, and they provide additional interpretations in the context of risk theory.

Another characterization of the increasing convex order, which will be used later, is the following.

Theorem 2.3.4. *Let X and Y be two random variables with finite means. Then, $X \leq_{\text{icx}} Y$ if, and only if,*

$$E[\max \{X, x\}] \leq E[\max \{Y, x\}], \quad \text{for all } x \in \mathbb{R}.$$

Proof. The proof follows observing that $(t - x)_+ = \max \{t, x\} - x$, and Theorem 2.3.2. □

From the definition of the increasing convex order, it is clear that

$$X \leq_{\text{st}} Y \Rightarrow X \leq_{\text{icx}} Y. \tag{2.6}$$

Therefore, the increasing convex order has interest not only in risk theory, but also in situations where the stochastic order does not hold. In fact, even if the survival functions cross each other just once, the increasing convex order can hold, as it occurs in the example considered in the introduction. Let us see in which situations it occurs.

Theorem 2.3.5. *Let X and Y be two random variables with survival functions $\overline{F}$ and $\overline{G}$, respectively, and finite means such that $E[X] \leq E[Y]$. If $S^- (\overline{G} - \overline{F}) = 1$ with the sign sequence $-, +$, then*

$$X \leq_{\text{icx}} Y.$$

Proof. In order to verify $X \leq_{\text{icx}} Y$, we shall prove that the condition provided in Theorem 2.3.4 holds. Let us assume that a value $x_0 \in \mathbb{R}$ exists such that $\overline{F}(x) \geq \overline{G}(x)$, for all $x \leq x_0$ and $\overline{F}(x) \leq \overline{G}(x)$, for all $x \geq x_0$. Then, it is easy to see that

$$\max \{X, x\} \leq_{\text{st}} \max \{Y, x\}, \quad \text{for all } x \geq x_0,$$

and, therefore, we see that

$$E[\max\{X, x\}] \leq E[\max\{Y, x\}], \quad \text{for all } x \geq x_0,$$

from (2.2). Let us consider now a value $x < x_0$. In this case, we see that

$$\min\{X, x\} \geq_{st} \min\{Y, x\}, \quad \text{for all } x \leq x_0,$$

and the proof follows analogously, observing that $\max\{t, x\} = t + x - \min\{t, x\}$ and $E[X] \leq E[Y]$. □

The example provided in the introduction satisfies the conditions stated in the last theorem and, therefore, it is an example where the increasing convex order holds but the stochastic order does not. This result is known as the cut-criterion or the one-crossing condition of Karlin and Novikoff [56], and it has been generalized to provide a sign change characterization of the increasing convex order [57]. An example where the previous theorem can be also applied is the case of two Pareto distributions.

Example 2.3.6. Let $X \sim P(a_1, k_1)$ and $Y \sim P(a_2, k_2)$ with survival functions $\overline{F}$ and $\overline{G}$, respectively. We assume $a_2, a_1 > 1$ in order to have finite means. According to Example 2.2.2, if $k_2 \leq k_1$ and $a_2 > a_1$, we see that $S^-(\overline{G}(x) - \overline{F}(x)) = 1$, for all $x \geq k_1$, with the sign sequence $-, +$. In addition, if

$$E[X] = \frac{a_1 k_1}{a_1 - 1} \leq \frac{a_2 k_2}{a_2 - 1} = E[Y],$$

we get $X \leq_{icx} Y$ from the previous theorem. We observe that the conditions $k_2 \leq k_1$ and $a_1 k_1/(a_1 - 1) \leq a_2 k_2/(a_2 - 1)$, when the equalities do not hold at the same time, imply $a_2 < a_1$. Therefore, to sum up, if $k_2 \leq k_1$ and $a_1 k_1/(a_1 - 1) \leq a_2 k_2/(a_2 - 1)$, when the equalities do not hold at the same time, then $X \leq_{icx} Y$, but $X \not\leq_{st} Y$ or $X \not\geq_{st} Y$. Figure 2.5 shows a particular example of this situation.

In a similar way to Theorem 2.2.6, it is possible to establish a set of sufficient conditions for the increasing convex order in terms of the density functions.

Theorem 2.3.7. *Let X and Y be two continuous random variables with density functions f and g, respectively, and finite means such that $E[X] \leq E[Y]$. If $S^-(g - f) = 2$ with the sign sequence $+, -, +$, then*

$$X \leq_{icx} Y.$$

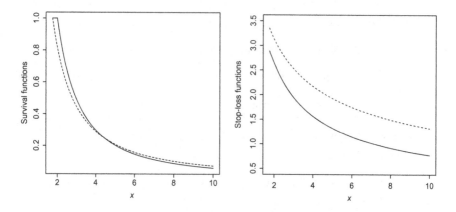

Figure 2.5 Survival functions on the left side and stop-loss functions on the right side of $X \sim P(1.75, 2)$ (continuous line) and $Y \sim P(1.5, 1.75)$ (dashed line).

Proof. This proof follows combining Theorem 2.3.5 with Corollary 1.2.4.   □

This result also provides another tool to compare two random variables in terms of the so-called ρ functions [58]. For a continuous random variable X with a differentiable density function f, the ρ_X function is defined as

$$\rho_X(x) = \frac{d}{dx} \log f(x), \quad \text{for all } x \in \mathbb{R} \text{ such that } f(x) > 0.$$

Let us state the theorem.

Theorem 2.3.8. *Let X and Y be two continuous random variables with a common interval support and finite means. If the functions ρ_X and ρ_Y are differentiable and $S^-(\rho_Y - \rho_X) = 1$ with the sign sequence $-, +$, then*

$S^-(g - f) \leq 2$ *with the sign sequence $+, -, +$ when the equality holds.*

Proof. Let us denote by x_0 the crossing point among ρ_Y and ρ_X. Observe that it is enough to prove that $S^-(g - f) \leq 1$ with the sign sequence $+, -$ when the equality holds, on the interval $(-\infty, x_0)$, and $S^-(g - f) \leq 1$ with the sign sequence $-, +$ when the equality holds, on the interval $(x_0, +\infty)$. Notice that $g - f$ cannot have a sign change $-, +$ on the interval $(-\infty, x_0)$, due to the fact that, if $x < x_0$ exists such that $g(x) = f(x)$, then

$$g(y) - f(y) = f(x) \exp\left\{ \int_x^y [\rho_Y(u) - \rho_X(u)] \, du \right\} \leq 0,$$

for all $x < y < x_0$. Analogously, we find that $g - f$ cannot have a sign change $+, -$ on the interval $(x_0, +\infty)$. Hence, $S^-(g-f) \leq 2$ with the sign sequence $+, -, +$ when the equality holds. $\qquad \square$

Observe that, from the previous theorem and Theorem 2.3.7, if $E[X] \leq E[Y]$ and

$$\lim_{x \to l} \frac{g(x)}{f(x)}, \lim_{x \to u} \frac{g(x)}{f(x)} > 1,$$

where l and u denote the left and right extremes of the common supports, respectively, then $X \leq_{\mathrm{icx}} Y$, but $X \not\leq_{\mathrm{st}} Y$ or $X \not\geq_{\mathrm{st}} Y$. Otherwise, the previous theorem implies $X \leq_{\mathrm{st}} Y$, since $S^-(g-f) = 1$ with the sign sequence $-, +$.

The following example shows the simplicity and usefulness of this result to compare, for instance, two gamma distributions.

Example 2.3.9. Let $X \sim G(\alpha_1, \beta_1)$ and $Y \sim G(\alpha_2, \beta_2)$. It is easy to see that $\rho_X(x) = \frac{\alpha_1 - 1}{x} - \frac{1}{\beta_1}$ and $\rho_Y(x) = \frac{\alpha_2 - 1}{x} - \frac{1}{\beta_2}$. Then $S^-(\rho_Y - \rho_X) = 1$, whenever the crossing point belongs to the support, which is equivalent to require

$$\frac{\alpha_2 - \alpha_1}{\frac{1}{\beta_2} - \frac{1}{\beta_1}} > 0.$$

Furthermore, the condition $\rho_Y(x) > \rho_X(x)$, for great values of x holds if $\alpha_2 < \alpha_1$ and $\beta_2 > \beta_1$. In addition, it is required that $E[X] = \alpha_1 \beta_1 \leq \alpha_2 \beta_2 = E[Y]$. To sum up, if $\alpha_2 < \alpha_1$ and $\alpha_1 \beta_1 \leq \alpha_2 \beta_2$, then $X \leq_{\mathrm{icx}} Y$, but $X \not\leq_{\mathrm{st}} Y$ or $X \not\geq_{\mathrm{st}} Y$, by the previous comment. Figure 2.6 shows a particular example of this situation.

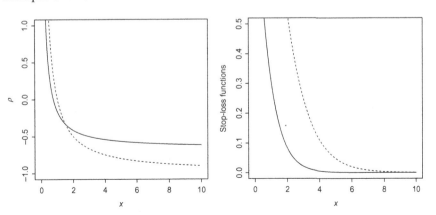

Figure 2.6 ρ functions on the left side and stop-loss functions on the right side of $X \sim G(2, 1)$ (continuous line) and $Y \sim G(1.5, 1.5)$ (dashed line).

Next, some preservation results for the increasing convex order are provided. First, the result on the preservation under increasing convex transformations is stated.

Theorem 2.3.10. *Let X and Y be two random variables. Then, $X \leq_{icx} Y$ if, and only if,*

$$\phi(X) \leq_{icx} \phi(Y),$$

for all real valued increasing convex function ϕ.

Proof. The proof follows observing that the composition of two increasing convex functions is also an increasing convex function. □

Next, a result on the preservation under convergence is provided.

Theorem 2.3.11. *Let $\{X_n\}_{n\in\mathbb{N}}$ and $\{Y_n\}_{n\in\mathbb{N}}$ be two sequences of random variables such that X_n converges in distribution to X and Y_n converges in distribution to Y. If $X_n \leq_{icx} Y_n$, for all $n \in \mathbb{N}$ and $E[(X_n)_+] \to E[(X)_+]$ and $E[(Y_n)_+] \to E[(Y)_+]$, then*

$$X \leq_{icx} Y.$$

Proof. Observe that

$$E[(X_n - x)_+] = E[(X_n)_+] - E[\min\{(X_n)_+, x\}], \quad \text{for all } x \in \mathbb{R},$$

and analogously to $E[(Y_n - x)_+]$. On the one hand, $E[(X_n)_+] \to E[(X)_+]$ and $E[(Y_n)_+] \to E[(Y)_+]$ by hypothesis. On the other hand, we have

$$E[\min\{(X_n)_+, x\}] \to E[\min\{(X)_+, x\}],$$

and analogously to $E[\min\{(Y_n)_+, x\}]$, since $\min\{t_+, x\}$ is a bounded continuous function and X_n $[Y_n]$ converges in distribution to X $[Y]$. Therefore,

$$E[(X_n - x)_+] \to E[(X - x)_+]$$

and analogously to $E[(Y_n - x)_+]$. Then, $E[(X - x)_+] \leq E[(Y - x)_+]$, since $X_n \leq_{icx} Y_n$. □

The following result copes with the preservation of the increasing convex order under mixtures.

Theorem 2.3.12. *Let $\{X_1(\theta)|\theta \in S \subseteq \mathbb{R}\}$ and $\{X_2(\theta)|\theta \in S \subseteq \mathbb{R}\}$ be two families of random variables, and Θ_1 and Θ_2 be two random variables with common support S. If*

$$X_1(\theta) \leq_{\text{icx}} X_2(\theta), \quad \textit{for all } \theta \in S, \tag{2.7}$$

$$\Theta_1 \leq_{\text{icx}} \Theta_2 \tag{2.8}$$

and $E[\phi(X_1(\theta))]$ or $E[\phi(X_2(\theta))]$ or both are increasing convex in θ, for all real valued increasing convex function ϕ, then

$$X_1(\Theta_1) \leq_{\text{icx}} X_2(\Theta_2).$$

Proof. Let us consider a real valued increasing convex function ϕ such that the expectations $E[\phi(X_1(\Theta_1))]$ and $E[\phi(X_2(\Theta_2))]$ exist, and let us suppose that $E[\phi(X_2(\theta))]$ is increasing convex in θ. Again, like in the proof of Theorem 2.2.8, we see that $E[\phi(X_1(\Theta_1))] = E[f_1(\Theta_1)]$ and $E[\phi(X_2(\Theta_2))] = E[f_2(\Theta_2)]$, where $f_1(\theta) = E[\phi(X_1(\theta))]$ and $f_2(\theta) = E[\phi(X_2(\theta))]$. Then,

$$E[\phi(X_1(\Theta_1))] = E[f_1(\Theta_1)] \leq E[f_2(\Theta_1)] \leq E[f_2(\Theta_2)] = E[\phi(X_2(\Theta_2))],$$

where the first inequality follows from (2.7) and the second one follows from (2.8) and the fact that $E[\phi(X_2(\theta))]$ is increasing convex in θ. The case where $E[\phi(X_1(\theta))]$ is increasing convex is similar. □

Additional results in this direction can be found in Ref. [59].

To finish, a result on the preservation under convolution of the increasing convex order is provided. As in the case of the stochastic order, the result follows by induction. In any case, the result follows from a general one in the multivariate case (see Theorem 3.3.6).

Theorem 2.3.13. *Let $X_1, \ldots, X_n$ and $Y_1, \ldots, Y_n$ be two sets of independent random variables. If $X_i \leq_{\text{icx}} Y_i$, for all $i = 1, \ldots, n$, then*

$$\sum_{i=1}^{n} X_i \leq_{\text{icx}} \sum_{i=1}^{n} Y_i.$$

Next, we provide a graphical tool to check the icx order for two samples by plotting their empirical stop-loss functions. Another possibility arises from the cut-criterion or one-crossing condition. If we plot the survival functions, it is enough to check if these functions cross each other once and, additionally, to verify if the means are ordered, which can be done by usual hypothesis tests.

Let us consider the following example for two data sets taken from Ref. [60] to illustrate the empirical validation of the icx order. In this study, the survival times of two groups of rats are recorded to see the effect on the longevity of a restricted diet (RD) versus an *ad libitum* diet (ALD), which means free eating. The plot on the left side of Figure 2.7 shows the empirical stop-loss functions and suggests the increasing convex order among the two groups, and the plot on the right side shows that the survival functions cross each other once. A t-student test confirms that the means are ordered and, therefore, there is empirical evidence for the increasing convex order among the two data sets.

To conclude this section, we consider two stochastic orders related to the increasing convex order.

In the case of two random variables X and Y such that $E[X] = E[Y]$, then $X \leq_{\text{icx}} Y$ if, and only if, $E[\phi(X)] \leq E[\phi(Y)]$, for all real valued convex function such that the previous expectations exists. That is, in case of equal means, the increasing convex order is not restricted to the comparison of increasing convex functions, it can be extended to convex functions. The inverse is also true, that is, if $E[\phi(X)] \leq E[\phi(Y)]$, for all real valued convex function such that the previous expectations exist, then $E[X] = E[Y]$ and $X \leq_{\text{icx}} Y$. In fact, when two random variables satisfy this condition, then it can be said that X is smaller than Y in the *convex order*, denoted by $X \leq_{\text{cx}} Y$.

If we replace increasing convex functions by increasing concave functions in the definition of the increasing convex order, we have the so-called increasing concave order. We say that X is smaller than Y in the *increasing concave order*, denoted by $X \leq_{\text{icv}} Y$, if $E[\phi(X)] \leq E[\phi(Y)]$ for all

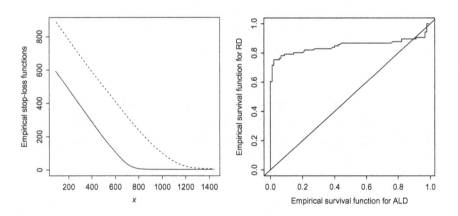

Figure 2.7 Empirical stop-loss functions for RD (continuous line) and ALD (dashed line) on the left side, and empirical $\bar{P} - \bar{P}$ plot for RD and ALD on the right side.

real valued increasing concave function such that the previous expectations exist. It is not difficult to see that

$$X \leq_{\text{icv}} Y \Leftrightarrow -Y \leq_{\text{icx}} -X. \tag{2.9}$$

The relationship among these two partial orders and the increasing convex order allows us to translate results on the increasing convex order to the convex and increasing concave ones.

2.4 THE HAZARD RATE AND MEAN RESIDUAL LIFE ORDERS

As observed previously, one way of involving in the comparison of the random variables as much information as possible is to compare their survival functions. Apart from the ones considered for the stochastic and the increasing convex orders, there are still many ways of establishing this comparison in terms of the survival functions. For example, another possibility arises from the ratio of the survival functions, by checking if $\overline{G}(x)/\overline{F}(x)$ is increasing or, equivalently,

$$\overline{F}(x)\overline{G}(y) \geq \overline{F}(y)\overline{G}(x), \quad \text{for all } x \leq y,$$

to avoid problems with zero values in the denominator. The interest of this property relies on the fact that the monotonicity of the ratio of the survival functions reflects that the survival function $\overline{F}$ decreases more quickly than $\overline{G}$ and, therefore, the random variable X tends to take on smaller values than Y. Therefore, it is natural to define another criterion through the previous condition. Let us see the formal definition of this criterion, which is named the hazard rate order.

Definition 2.4.1. Given two random variables X and Y with survival functions $\overline{F}$ and $\overline{G}$, respectively, we say that X is smaller than Y in the *hazard rate order*, denoted by $X \leq_{\text{hr}} Y$, if

$$\overline{F}(x)\overline{G}(y) \geq \overline{F}(y)\overline{G}(x), \quad \text{for all } x \leq y, \tag{2.10}$$

or, equivalently, if $\overline{G}(x)/\overline{F}(x)$ is increasing in $\{x|\overline{F}(x) > 0\}$.

The hazard rate order is stronger than the stochastic order, that is,

$$X \leq_{\text{hr}} Y \Rightarrow X \leq_{\text{st}} Y. \tag{2.11}$$

The proof follows by taking limits as x tends to $-\infty$ in (2.10).

As in the case of the stochastic order, the simplicity of this criterion often allows us to check it directly from its definition. For instance, this is the case of two Pareto distributions.

Example 2.4.2. Let $X \sim P(a_1, k_1)$ and $Y \sim P(a_2, k_2)$ with survival functions $\overline{F}$ and $\overline{G}$, respectively. Denote by $H(x) = \overline{G}(x)/\overline{F}(x)$. On the one hand, we see that $H(x) = \left(\frac{k_2^{a_2}}{k_1^{a_1}} \right) x^{a_1 - a_2}$, for all $x \geq \max\{k_1, k_2\}$, which is increasing in x when $a_1 \geq a_2$. On the other hand, we have

$$H(x) = \begin{cases} \frac{1}{\overline{F}(x)} & \text{if } k_1 \leq k_2, \\ \\ \overline{G}(x) & \text{if } k_2 < k_1, \end{cases}$$

for all $x \leq \max\{k_1, k_2\}$ and, consequently, $H(x)$ is decreasing if $k_2 < k_1$ and increasing if $k_1 \leq k_2$, for all $x \leq \max\{k_1, k_2\}$. In conclusion, if $a_1 \geq a_2$ and $k_1 \leq k_2$, it holds $X \leq_{\mathrm{hr}} Y$. Therefore, we have a stronger conclusion than the one provided in Example 2.2.2, since the hazard rate order holds under the same conditions on the parameters obtained for the stochastic order. Figure 2.8 shows a particular example of this situation.

If the distribution functions are differentiable, the hazard rate order can be trivially characterized in terms of the hazard rate functions. In particular,

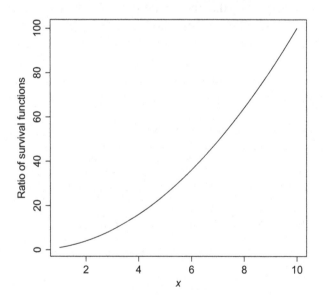

Figure 2.8 Ratio of survival functions of $X \sim P(4, 1)$ and $Y \sim P(2, 1)$.

denoting by r and s the hazard rates of X and Y, respectively, then $X \leq_{\mathrm{hr}} Y$ if, and only if,

$$r(x) \geq s(x), \quad \text{for all } x \text{ such that } \overline{F}(x), \overline{G}(x) > 0.$$

The hazard rate order can be also related to the comparison of the residual lives, which provide an application not only in reliability, but also in risk theory. From (1.4) and (2.10), it is not difficult to prove the following result.

Theorem 2.4.3. *Let X and Y be two random variables with survival functions $\overline{F}(x)$ and $\overline{G}(x)$, respectively. Then, $X \leq_{\mathrm{hr}} Y$ if, and only if,*

$$[X - x | X > x] \leq_{\mathrm{st}} [Y - x | Y > x], \quad \text{for all } x \text{ such that } \overline{F}(x), \overline{G}(x) > 0.$$

Next, some preservation results are provided. First, a result on preservation under convergence is given.

Theorem 2.4.4. *Let $\{X_n\}_{n \in \mathbb{N}}$ and $\{Y_n\}_{n \in \mathbb{N}}$ be two sequences of random variables such that X_n converges in distribution to X and Y_n converges in distribution to Y. If $X_n \leq_{\mathrm{hr}} Y_n$, for all $n \in \mathbb{N}$, then*

$$X \leq_{\mathrm{hr}} Y.$$

Proof. The proof is similar to the one of Theorem 2.2.7, taking into account (2.10). □

The hazard rate order is also preserved under increasing transformations, as we shall see next.

Theorem 2.4.5. *Let X and Y be two random variables. If $X \leq_{\mathrm{hr}} Y$, then*

$$\phi(X) \leq_{\mathrm{hr}} \phi(Y), \quad \text{for all real valued increasing function } \phi.$$

Proof. Let us consider first that ϕ is a strictly increasing function. It is not difficult to see that $\{\phi(X) \leq x\} = \{X \leq \phi^{-1}(x)\}$, for all $x \in \mathbb{R}$ and analogously for $\phi(Y)$. Therefore, $P[\phi(X) > x] = \overline{F}(\phi^{-1}(x))$ and $P[\phi(Y) > x] = \overline{G}(\phi^{-1}(x))$. The proof follows from (2.10) and the increase of ϕ. If ϕ is just increasing, the result follows from Theorem 2.4.4, since every increasing function can be approximated by a sequence of strictly increasing functions. □

Now, the preservation of the hazard rate order under mixtures is considered. In this case, there is no similar result on the preservation of the hazard rate order under mixtures to those provided for the stochastic order and the increasing convex order. There is a result on the comparison of just one family of mixing distributions.

Theorem 2.4.6. *Let $\{X(\theta), \, \theta \in S \subseteq \mathbb{R}\}$ be a family of random variables, and Θ_1 and Θ_2 be two random variables with common support S. If*

$$X(\theta) \leq_{hr} X(\theta'), \quad \text{for all } \theta \leq \theta' \in S, \tag{2.12}$$

and

$$\Theta_1 \leq_{hr} \Theta_2, \tag{2.13}$$

then

$$X(\Theta_1) \leq_{hr} X(\Theta_2).$$

Proof. First, let us fix some notation. The distribution function of $X(\theta)$ will be denoted by $F_\theta(x)$, and the distribution function of Θ_i will be denoted by $H_i(\theta)$, for $i = 1, 2$. Let $x \leq y$ and denote $l_1(\theta) = \overline{F}_\theta(x)$ and $l_2(\theta) = \overline{F}_\theta(y)$. On the one hand, from (2.12), we see that

$$l_1(\theta)l_2(\theta') \geq l_1(\theta')l_2(\theta),$$

and, from (2.11), we also have that $l_1(\theta)$ is increasing in θ.

On the other hand, (2.13) is equivalent to $\overline{H}_i(\theta)$ be TP$_2$ in $\{1, 2\} \times S$.

Therefore, applying Theorem 1.2.6, we get

$$\int \overline{F}_\theta(x) dH_1(\theta) \int \overline{F}_\theta(y) dH_2(\theta) \geq \int \overline{F}_\theta(x) dH_2(\theta) \int \overline{F}_\theta(y) dH_1(\theta),$$

for all $x \leq y$, that is, $X(\Theta_1) \leq_{hr} X(\Theta_2)$. $\qquad\square$

Next, several results on the preservation of the hazard rate order under convolutions are introduced.

Theorem 2.4.7. *Let X, Y and Z be three random variables such that Z is independent of X and Y. If $X \leq_{hr} Y$, and Z is IFR, then*

$$X + Z \leq_{hr} Y + Z.$$

Proof. Let us consider the random variables $[X + Z | X = x]$ and $[Y + Z | Y = x]$, for all x. From the independence hypothesis, they are equally distributed and it is not difficult to see that

$$[X + Z | X = x] \leq_{hr} [Y + Z | Y = x'], \quad \text{for all } x \leq x',$$

since Z is IFR. Therefore, if we consider the mixture of $[X + Z | X = x]$ and $[Y + Z | Y = x]$, we find that $X + Z \leq_{hr} Y + Z$, from Theorem 2.4.6. $\qquad\square$

Repeatedly, the application of this theorem and the preservation by the convolution of IFR distributions (see Theorem 1.2.1) leads to the following general result.

Theorem 2.4.8. *Let* $\{(X_i, Y_i)\}_{i=1}^{n}$, *be independent pairs of random variables. If* X_i *and* Y_i *are IFR and* $X_i \leq_{\mathrm{hr}} Y_i$, *for all* $i = 1, \ldots, n$, *then*

$$\sum_{i=1}^{n} X_i \leq_{\mathrm{hr}} \sum_{i=1}^{n} Y_i.$$

Proof. From the previous theorem, we see that

$$X_1 + X_2 \leq_{\mathrm{hr}} X_1 + Y_2 \leq_{\mathrm{hr}} Y_1 + Y_2.$$

Repeating this argument, since the IFR notion is preserved under convolutions (see Theorem 1.2.1), we get the result.   □

If the survival functions are replaced by the distribution functions in (2.10), we get the definition of the *reversed hazard rate order*, denoted by $X \leq_{\mathrm{rh}} Y$. The hazard rate is related to the reversed hazard rate order in the following way. If $X \leq_{\mathrm{rh}} Y$, then $\phi(Y) \leq_{\mathrm{hr}} \phi(X)$, for all continuous strictly decreasing function ϕ [61] and, therefore, $-X \geq_{\mathrm{hr}} -Y$. Taking into account this comment, most of the results for the hazard rate order can be translated to the reversed hazard rate order.

Similarly to the extension of the stochastic order to the increasing convex order, from both the characterization provided in (2.10) and Theorem 2.4.3 of the hazard rate order, it is natural to consider the following conditions as possible candidates to keep comparing random quantities in terms of their residual lives

$$\int_{x}^{+\infty} \overline{F}(u) \, du \int_{y}^{+\infty} \overline{G}(u) \, du \geq \int_{y}^{+\infty} \overline{F}(u) \, du \int_{x}^{+\infty} \overline{G}(u) \, du,$$

for all $x \leq y$, and

$$[X - x | X > x] \leq_{\mathrm{icx}} [Y - x | Y > x],$$

for all x such that $\overline{F}(x), \overline{G}(x) > 0$. In the continuous case, these conditions are related to the comparison of the mean residual life functions—that is, they are equivalent to

$$m(x) \leq l(x), \quad \text{for all } x \text{ such that } \overline{F}(x), \overline{G}(x) > 0, \tag{2.14}$$

where m and l denote the mean residual life functions of X and Y, respectively. The criterion we are referring to is the mean residual life order and it is defined through the condition (2.14). The upcoming definition holds for continuous random variables, the case of discrete random variables should be modified [2, pp. 82–83].

Definition 2.4.9. Given two continuous random variables X and Y with survival functions $\overline{F}$ and $\overline{G}$, respectively, we say that X is smaller than Y in the *mean residual life order*, denoted by $X \leq_{\mathrm{mrl}} Y$, if

$$m(x) \leq l(x), \quad \text{for all } x \text{ such that } \overline{F}(x), \overline{G}(x) > 0.$$

Recall that the mean residual life function exists if the mean is finite, hence this condition will be assumed in the remaining part of this section.

The mean residual life order can be characterized as follows, as we have already pointed out.

Theorem 2.4.10. *Let X and Y be two continuous random variables with survival functions $\overline{F}$ and $\overline{G}$, respectively. Then, $X \leq_{\mathrm{mrl}} Y$ if, and only if,*

$$\int_x^{+\infty} \overline{F}(u)\, du \int_y^{+\infty} \overline{G}(u)\, du \geq \int_y^{+\infty} \overline{F}(u)\, du \int_x^{+\infty} \overline{G}(u)\, du, \quad (2.15)$$

for all $x \leq y$ or, equivalently,

$$[X - x|X > x] \leq_{\mathrm{icx}} [Y - x|Y > x], \quad (2.16)$$

for all x such that $\overline{F}(x), \overline{G}(x) > 0$.

Proof. Let us see first the equivalence with (2.15). The inequality (2.15) is equivalent to

$$\frac{\int_x^{+\infty} \overline{G}(u)\, du}{\int_x^{+\infty} \overline{F}(u)\, du} \text{ be increasing in } x \text{ for all } \{x|\overline{F}(x) > 0\}, \quad (2.17)$$

and, taking derivatives on the previous expression with respect to x, we get the result.

Let us see now (2.16). Notice that (2.16) holds if, and only if,

$$\frac{\int_y^{+\infty} \overline{F}(u)\, du}{\overline{F}(x)} \leq \frac{\int_y^{+\infty} \overline{G}(u)\, du}{\overline{G}(x)}, \quad \text{for all } x \leq y \text{ such that } \overline{F}(x), \overline{G}(x) > 0,$$

therefore (2.16) implies the mean residual life order. Next, the inverse implication is proved. Let us suppose that $X \leq_{\mathrm{mrl}} Y$. From (2.14) and (2.17), we have the following chain of inequalities:

$$\frac{\overline{G}(x)}{\overline{F}(x)} \leq \frac{\int_x^{+\infty} \overline{G}(u)\, du}{\int_x^{+\infty} \overline{F}(u)\, du} \leq \frac{\int_y^{+\infty} \overline{G}(u)\, du}{\int_y^{+\infty} \overline{F}(u)\, du}, \qquad (2.18)$$

for all $x \leq y$ such that $\overline{F}(x), \overline{G}(x) > 0$, which concludes the proof. $\square$

Clearly, from the definition of the mean residual life order, Theorem 2.4.3 and (2.2), we have

$$X \leq_{\mathrm{hr}} Y \Rightarrow X \leq_{\mathrm{mrl}} Y.$$

In addition, from (2.18), we have

$$X \leq_{\mathrm{mrl}} Y \Rightarrow X \leq_{\mathrm{icx}} Y.$$

According to we have seen up to this point, notice that the verification of the mean residual life requires the evaluation of the incomplete integrals of the survival functions. Unfortunately, there are many situations where this is not possible. Next, we introduce some sufficient conditions for the mean residual life order that allow us to overcome this difficulty (see Ref. [62] for a discussion on the topic).

Theorem 2.4.11. *Let X and Y be two continuous random variables with survival functions $\overline{F}$ and $\overline{G}$, respectively, and finite means such that $E[X] \leq E[Y]$. If a value $x_0 \in \mathbb{R}$ exists such that $\overline{F}(x)\overline{G}(y) \leq \overline{F}(y)\overline{G}(x)$, for all $x < y \leq x_0$ and $\overline{F}(x)\overline{G}(y) \geq \overline{F}(y)\overline{G}(x)$, for all $x_0 \leq x < y$, then*

$$X \leq_{\mathrm{mrl}} Y.$$

Proof. From the assumption, we see that $\overline{F}(x)\overline{G}(y) \geq \overline{F}(y)\overline{G}(x)$, for all $x_0 \leq x < y$, and integrating with respect to y we get

$$\overline{F}(x) \int_x^{+\infty} \overline{G}(y)\, dy \geq \overline{G}(x) \int_x^{+\infty} \overline{F}(y)\, dy, \quad \text{for all } x > x_0. \qquad (2.19)$$

Now, let $x \leq y \leq x_0$, then $\overline{F}(x)\overline{G}(y) \leq \overline{F}(y)\overline{G}(x)$ and taking limits as x tends to $-\infty$, we obtain

$$\overline{F}(y) \geq \overline{G}(y), \quad \text{for all } y \leq x_0. \qquad (2.20)$$

This inequality is equivalent to

$$\min\{X, y\} \geq_{\mathrm{st}} \min\{Y, y\}, \quad \text{for all } y \leq x_0.$$

Therefore, from (2.2), we have

$$E[\min\{X, y\}] \geq E[\min\{Y, y\}], \quad \text{for all } y \leq x_0.$$

Replacing $\min\{X, y\} = X - (X - y)_+$ (and analogously for Y) in the previous expression, since $E[X] \leq E[Y]$, we get

$$E[(X - y)_+] \leq E[(Y - y)_+], \quad \text{for all } y \leq x_0.$$

Combining this inequality with (2.20) we conclude

$$\overline{G}(y)E[(X - y)_+] \leq \overline{F}(y)E[(Y - y)_+], \quad \text{for all } y \leq x_0. \tag{2.21}$$

The result follows from (2.19) and (2.21).   □

The hypothesis of the previous theorem can be rewritten in terms of the ratio $H(x) = \overline{G}(x)/\overline{F}(x)$. In particular, the assumption holds if, and only if, $H(x)$ has just one minimum. If a function h has just one extreme, we shall say that h is unimodal. Next, we provide an example where the previous result is applied to compare two Pareto distributions in the mean residual life order.

Example 2.4.12. Let $X \sim P(a_1, k_1)$ and $Y \sim P(a_2, k_2)$ with survival functions $\overline{F}$ and $\overline{G}$, respectively. We assume $a_1, a_2 \geq 1$ to have finite means. From Example 2.3.6, if $k_2 < k_1$ and $E[X] = a_1k_1/(a_1 - 1) \leq E[Y] = a_2k_2/(a_2 - 1)$, then $a_2 < a_1$. Recalling now the discussion given in Example 2.4.2, if $k_2 < k_1$ and $a_2 < a_1$, then the ratio of the survival functions is initially deceasing and later increasing. Therefore, from the previous theorem, if $k_2 < k_1$ and

$$\frac{a_1k_1}{a_1 - 1} \leq \frac{a_2k_2}{a_2 - 1},$$

then $X \leq_{\mathrm{mrl}} Y$, but $X \not\leq_{\mathrm{hr}} Y$ or $X \not\geq_{\mathrm{hr}} Y$. Figure 2.9 shows a particular example of this situation.

The unimodality of the ratio of the survival functions has been already considered by Metzger and Rüschendorf [63]. In their Proposition 2.2.b, they relate the unimodality to crossing hazard rates functions. In particular, they proved the following. Let X and Y be two continuous random variables with survival functions $\overline{F}$ and $\overline{G}$, respectively. Then, $\overline{G}(x)/\overline{F}(x)$ is initially decreasing and later increasing if, and only if, a value $x_0 \in \mathbb{R}$ exists such that $r(x) \leq s(x)$, for all $x \leq x_0$ and $r(x) \geq s(x)$, for all $x \geq x_0$. Therefore, Theorem 2.4.11 can be written as follows.

Theorem 2.4.13. *Let X and Y be two continuous random variables with hazard rates functions r and s, respectively and finite means such that $E[X] \leq E[Y]$. If $S^-(s - r) = 1$ with the sign sequence $+, -$ on the set $\{x \in \mathbb{R}|\overline{G}(x), \overline{F}(x) > 0\}$, then*

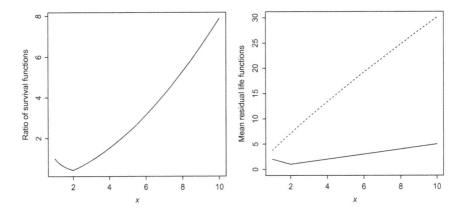

Figure 2.9 Ratio of survival functions of $X \sim P(3,2)$ and $Y \sim P(1.2,1)$ on the left side, and mean residual life functions on the right side of $X \sim P(3,2)$ (continuous line) and $Y \sim P(1.2,1)$ (dashed line).

$$X \leq_{\text{mrl}} Y.$$

Crossing hazard rate functions is a topic of interest in survival analysis. Let consider the following situation. A treatment is tested by comparing the survival times of a group of illness people versus a control group. In such a situation, if the hazard rate functions cross each other once, then the treatment has benefits only in the early stage of a disease but does not have long-term advantages, or the treatment has benefits in the long run, but it may increase the risk in the early stage of the disease. Theorem 2.4.13 shows that, if the means are ordered, the mean residual life order holds in those situations where the hazard rate functions cross each other once. Therefore, it is possible to keep comparing in some sense the residual lifetimes. Next, we apply the previous result to compare two Weibull distributions.

Example 2.4.14. Let $X \sim W(\alpha_1, \beta_1)$ and $Y \sim W(\alpha_2, \beta_2)$ with survival functions $\overline{F}$ and $\overline{G}$, respectively. The discussion will be given in terms of crossing hazard rates, according to the previous theorem. The hazard rates of X and Y will be denoted by r and s, respectively. Let us consider the following cases:

(a) $\beta_1 > \beta_2 \geq 1$. In this case, the hazard rates are increasing and $r(t) \geq s(t)$, for great values of t.

(b) $\beta_1 > 1 > \beta_2 \geq 1$. In this case, $r(t)$ is increasing and $s(t)$ is decreasing so, obviously, $r(t) \geq s(t)$, for great values of t.

(c) $1 \geq \beta_1 > \beta_2$. In this case, the hazard rates are decreasing and $r(t) \geq s(t)$, for great values of t.

Therefore, if $\beta_1 > \beta_2$, we have $S^-(s - r) = 1$ with the sign sequence $+, -$ on the set $\{x \in \mathbb{R} | \overline{G}(x), \overline{F}(x) > 0\}$. To sum up, if $\beta_1 > \beta_2$ and

$$E[X] = \alpha_1 \Gamma(1 + 1/\beta_1) \leq \alpha_2 \Gamma(1 + 1/\beta_2) = E[Y],$$

then $X \leq_{mrl} Y$, but $X \not\leq_{hr} Y$ or $X \not\geq_{hr} Y$, by the previous theorem. Figure 2.10 shows a particular example of this situation.

It is worth pointing out the relevance of this example in reliability, since it is widely known that the hazard rate order never holds for any pair of Weibull distributions (except that they have the same shape parameter) and it was uncertain if they can be ordered in the mean residual life order so far. This example makes clear that there is a great deal of cases where the mean residual life order holds among two Weibull distributions.

Although the conditions of Theorem 2.4.13 are satisfied for numerous random variables with no explicit expression for the mean residual functions, there are still many uncovered situations. One of the reasons why this tool is not enough is because of the fact that the survival functions may not have closed expressions either. For instance, the gamma or normal distributions present this behavior. According to Theorem 2.3 by Metzger and Rüschendorf [63], it is possible to give sufficient conditions in terms of the density functions as well. They prove that if the ratio of the density

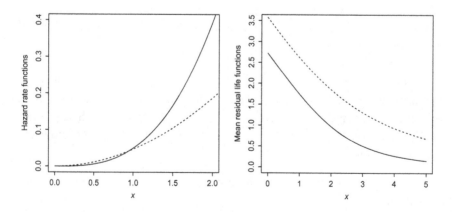

Figure 2.10 Hazard rate functions on the left side and mean residual life functions on the right side of $X \sim W(3, 4)$ (continuous line) and $Y \sim W(4, 3)$ (dashed line).

functions is unimodal, then the ratio of the survival functions has the same behavior. Combining this fact with Theorem 2.4.11, the following result is obtained.

Theorem 2.4.15. *Let X and Y be two continuous random variables with density functions f and g, respectively, with common support and finite means such that $E[X] \leq E[Y]$. If a value x_0 exists such that $g(x)/f(x)$ is decreasing in $x \leq x_0$, and $g(x)/f(x)$ is increasing in $x \geq x_0$, then*

$$X \leq_{\text{mrl}} Y.$$

Let us apply this result to compare two normal and gamma distributions.

Example 2.4.16. Let $X \sim N(\mu_1, \sigma_1^2)$ and $Y \sim N(\mu_2, \sigma_2^2)$ with density functions f and g, respectively. The behavior of the function

$$\frac{g(x)}{f(x)} = \frac{\sigma_1}{\sigma_2} \exp\left\{-\frac{1}{2\sigma_2^2}(x - \mu_2)^2 + \frac{1}{2\sigma_1^2}(x - \mu_1)^2\right\}, \quad \text{for all } x \in \mathbb{R},$$

is equivalent to the behavior of

$$h(x) = \frac{(x - \mu_1)^2}{\sigma_1^2} - \frac{(x - \mu_2)^2}{\sigma_2^2}, \quad \text{for all } x \in \mathbb{R}. \tag{2.22}$$

This function is unimodal and attains its extrema at

$$x_0 = \frac{\frac{\mu_2}{\sigma_2^2} - \frac{\mu_1}{\sigma_1^2}}{\frac{1}{\sigma_2^2} - \frac{1}{\sigma_1^2}}.$$

Taking limits as x tends to $\pm\infty$ in (2.22), we see that x_0 is a minimum if, and only if, $\sigma_1 < \sigma_2$. If we also assume $\mu_1 \leq \mu_2$, then $X \leq_{\text{mrl}} Y$, but $X \nleq_{\text{hr}} Y$ and $X \ngeq_{\text{hr}} Y$, by Theorem 2.4.15.

Example 2.4.17. Let $X \sim G(\alpha_1, \beta_1)$ and $Y \sim G(\alpha_2, \beta_2)$ with density functions f and g, respectively. The behavior of the function

$$\frac{g(x)}{f(x)} = \frac{\beta_1^{\alpha_1} \Gamma(\alpha_1)}{\beta_2^{\alpha_2} \Gamma(\alpha_2)} x^{\alpha_2 - \alpha_1} \exp\left\{\frac{x}{\beta_1} - \frac{x}{\beta_2}\right\}, \quad \text{for all } x > 0,$$

is equivalent to the behavior of

$$h(x) = (\alpha_2 - \alpha_1) \log(x) - \frac{x}{\beta_2} + \frac{x}{\beta_1}, \quad \text{for all } x > 0.$$

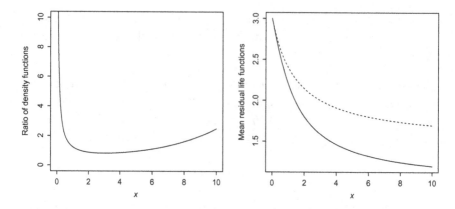

Figure 2.11 *Ratio of the density functions of $X \sim G(3, 1)$ and $Y \sim G(2, 3/2)$ on the left side, and mean residual life functions of $X \sim G(3, 1)$ (continuous line) and $Y \sim G(2, 3/2)$ (dashed line) on the right side.*

It is easy to see that $h(x)$ attains its extrema at

$$x_0 = \frac{\alpha_2 - \alpha_1}{\frac{1}{\beta_2} - \frac{1}{\beta_1}}.$$

Taking limits as x tends to $+\infty$ and 0 in $h(x)$, we see that x_0 is a minimum if, and only if, $\alpha_1 > \alpha_2$ and $\beta_2 > \beta_1$. If we also assume $E[X] = \alpha_1 \beta_1 \leq E[Y] = \alpha_2 \beta_2$, then $X \leq_{\mathrm{mrl}} Y$, but $X \not\leq_{\mathrm{hr}} Y$ and $X \not\geq_{\mathrm{hr}} Y$, by Theorem 2.4.15. Figure 2.11 shows a particular example of this situation.

If we consider a strictly increasing transformation of the random variables, $\phi(X)$ and $\phi(Y)$, we have $P[\phi(X) > x] = \overline{F}(\phi^{-1}(x))$ and $P[\phi(Y) > x] = \overline{G}(\phi^{-1}(x))$ and the behavior of the ratio $P[\phi(Y) > x]/P[\phi(X) > x]$ is the same as the behavior of $\overline{G}(x)/\overline{F}(x)$. If ϕ is just increasing, the result follows from the approximation of ϕ by strictly increasing functions.

From the previous comment and Example 2.4.16, two lognormal distributions can be compared in the mean residual life order.

Example 2.4.18. Let $X \sim LN(\mu_1, \sigma_1^2)$ and $Y \sim LN(\mu_2, \sigma_2^2)$. Notice that $X = \exp\{X'\}$ and $Y = \exp\{Y'\}$, where $X' \sim N(\mu_1, \sigma_1^2)$ and $Y' \sim N(\mu_2, \sigma_2^2)$, respectively. Then, if $\mu_1 \leq \mu_2$ and $\sigma_1 < \sigma_2$, we have $X \leq_{\mathrm{mrl}} Y$, $X \not\leq_{\mathrm{hr}} Y$ and $X \not\geq_{\mathrm{hr}} Y$. Notice that, under the previous conditions, $E[X] \leq E[Y]$ is also satisfied.

Next, some preservation results are provided. As we shall see, some of the results, and the arguments to prove these results, are closely related to those provided for the icx and hr orders.

Theorem 2.4.19. *Let* $\{X_n\}_{n\in\mathbb{N}}$ *and* $\{Y_n\}_{n\in\mathbb{N}}$ *be two sequences of random variables such that* X_n *converges in distribution to* X *and* Y_n *converges in distribution to* Y. *If* $X_n \leq_{\mathrm{mrl}} Y_n$, *for all* $n \in \mathbb{N}$ *and* $E[(X_n)_+] \to E[(X)_+]$ *and* $E[(Y_n)_+] \to E[(Y)_+]$, *then*

$$X \leq_{\mathrm{mrl}} Y.$$

Proof. The proof follows with analogous arguments to the ones in the proof of Theorem 2.3.11 and taking into account (2.15). □

Next, the preservation of the mrl order under increasing convex transformations is given.

Theorem 2.4.20. *Let* X *and* Y *be two random variables. Then,* $X \leq_{\mathrm{mrl}} Y$ *if, and only if,*

$$\phi(X) \leq_{\mathrm{mrl}} \phi(Y), \text{ for all real valued increasing convex function } \phi.$$

Proof. First, we observe that every increasing convex function ϕ is certainly strictly increasing and, therefore, as in the proof of Theorem 2.4.5, we have $\{\phi(X) \leq x\} = \{X \leq \phi^{-1}(x)\}$ and $\{\phi(Y) \leq x\} = \{Y \leq \phi^{-1}(x)\}$, for all $x \in \mathbb{R}$. Besides, notice that (2.16) can be replaced by

$$[X|X > x] \leq_{\mathrm{icx}} [Y|Y > x], \quad \text{for all } x \text{ such that } \overline{F}(x), \overline{G}(x) > 0,$$

and, consequently, from Theorem 2.3.10, we have

$$[\phi(X)|\phi(X) > x] \leq_{\mathrm{icx}} [\phi(Y)|\phi(Y) > x],$$

for all x such that $P[\phi(X) > x], P[\phi(Y) > x] > 0$. □

Next, a result on the comparison of mixtures in the mean residual life order is established.

Theorem 2.4.21. *Let* $\{X(\theta), \theta \in S\}$ *be a family of random variables, and* Θ_1 *and* Θ_2 *be two random variables with common support* S. *If*

$$X(\theta) \leq_{\mathrm{mrl}} X(\theta'), \quad \text{for all } \theta \leq \theta' \in S,$$

and

$$\Theta_1 \leq_{\mathrm{hr}} \Theta_2,$$

then

$$X(\Theta_1) \leq_{\mathrm{mrl}} X(\Theta_2).$$

Proof. According to the notation in Theorem 2.4.6, let us consider the functions $l_1(\theta) = \int_x^{+\infty} \overline{F}_\theta(u)\,du$ and $l_2(\theta) = \int_y^{+\infty} \overline{F}_\theta(u)\,du$, for all $x \leq y$. The proof follows under similar steps to those in the proof of Theorem 2.4.6.

□

Next, some results on the preservation under convolutions of the mean residual life order are provided. The proof of the following result is omitted; it can be found in Ref. [2].

Theorem 2.4.22. *Let X, Y and Z be three random variables such that Z is independent of X and Y. If $X \leq_{mrl} Y$, and Z is IFR, then*

$$X + Z \leq_{mrl} Y + Z.$$

Analogously to Theorem 2.4.8, repeated application of this theorem leads to the following general result.

Theorem 2.4.23. *Let $\{(X_i, Y_i)\}_{i=1}^n$ be independent pairs of random variables. If X_i and Y_i are IFR and $X_i \leq_{mrl} Y_i$, for all $i = 1,\ldots,n$, then*

$$\sum_{i=1}^n X_i \leq_{mrl} \sum_{i=1}^n Y_i.$$

Finally, according to the organization of previous sections, a graphical procedure to check the hazard rate and the mean residual life orders are given. First, we focus on the hazard rate order, which can be checked through the $\overline{P} - \overline{P}$ plot. First, we recall the definition of a star-shaped function. A subset A of the Euclidean space is called star-shaped with respect to a point s, if for all $x \in A$, then A contains the whole line segment between x and s. A real function f is called star-shaped with respect to a point (a, b), if its epigraph is star-shaped with respect to (a, b). According to this notion, it is easy to see that $X \leq_{hr} Y$ if, and only if, the $\overline{P} - \overline{P}$ plot is star-shaped with respect to $(0, 0)$. Let us see an application of this procedure to a data set. If we observe Figure 2.4, the empirical $\overline{P} - \overline{P}$ plot suggests that the hazard rate order is satisfied. Observe also that a convex $\overline{P} - \overline{P}$ plot implies this condition.

Next, we focus on the mean residual life order. In this case, there are two possibilities to check this criterion graphically. On the one hand, it is possible to verify this order through the $\overline{P} - \overline{P}$ plot by means of the sufficient conditions given by Theorem 2.4.11. In particular, $\overline{G}(x)/\overline{F}(x)$ is initially decreasing and later increasing if, and only if, a value $x_0 \in \mathbb{R}$ exists

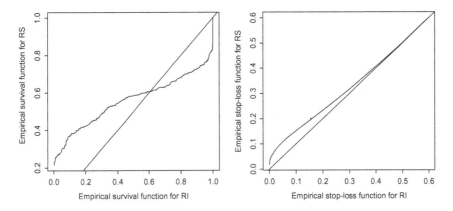

Figure 2.12 Empirical $\overline{P} - \overline{P}$ plot for RI and RS on the left side, and empirical stop-loss functions for RI and RS on the right side.

such that $\overline{P} - \overline{P}$ plot is anti star-shaped with respect to $(0,0)$ from $(0,0)$ until $(\overline{F}(x_0), \overline{G}(x_0))$, and star-shaped with respect to $(0,0)$ from $(\overline{F}(x_0), \overline{G}(x_0))$ up to $(1,1)$. If this condition is verified, the mean residual life order holds whenever the means are ordered. On the other hand, it is possible to plot the stop-loss functions against each other. If the epigraph of this plot is star-shaped with respect to $(0,0)$, we have, repeatedly, that the ratio of the stop loss functions is increasing in x and, therefore, by (2.17), the mean residual life order holds. Let us consider the following data set. We consider the weekly returns of a Spanish energetic company, Iberdrola (RI), and the weekly returns of a Spanish bank company, Santander (RS). The data have been obtained from the Yahoo! Finance web address. We have taken samples of size 500. Figure 2.12 shows the empirical $\overline{P} - \overline{P}$ plot of the two samples (on the left side) and the empirical stop-loss functions against each other. The plot suggests the mean residual life order among the two data sets.

2.5 THE LIKELIHOOD RATIO ORDER

As mentioned in previous sections, there are many random variables that do not have explicit expression for their distribution function. Therefore, it is difficult to check the hazard rate order in these situations. In this case, there is also a sufficient condition in terms of the density functions. The sufficient condition is another criterion, which is called the likelihood ratio order. This criterion can be also defined for discrete random variables, but we shall just focus on the case of continuous random variables. Although we have mainly introduced the likelihood ratio order as a sufficient condition for the hazard rate order, we recall several properties for the likelihood ratio order, in the same spirit as in previous sections. Let us see the formal definition.

Definition 2.5.1. Given two continuous random variables X and Y with density functions f and g, respectively, we say that X is smaller than Y in the *likelihood ratio order*, denoted by $X \leq_{lr} Y$, if

$$f(x)g(y) \geq f(y)g(x), \qquad \text{for all } x \leq y,$$

or equivalently, if

$\dfrac{g(x)}{f(x)}$ is increasing in x over the union of the supports of X and Y.

Notice that the monotonicity of the ratio of the density functions implies that $S^-(g - f) = 1$ and, therefore, applying Theorem 2.2.6, we see that

$$X \leq_{lr} Y \Rightarrow X \leq_{st} Y. \tag{2.23}$$

Moreover, the likelihood ratio order is stronger than the hazard rate order, as we have already pointed out. This statement can be obtained as a direct consequence of the following characterization.

Theorem 2.5.2. *Let X and Y be two continuous random variables. Then, $X \leq_{lr} Y$ if, and only if,*

$$[X | x \leq X \leq y] \leq_{st} [Y | x \leq Y \leq y], \tag{2.24}$$

for all $x \leq y$ such that $P[x \leq X \leq y], P[x \leq Y \leq y] > 0$.

Proof. Assuming that (2.24) holds, we find that

$$\frac{P[x \leq X \leq u]}{P[x \leq Y \leq u]} \geq \frac{P[u \leq X \leq y]}{P[u \leq Y \leq y]},$$

and

$$\frac{P[u \leq X \leq y]}{P[u \leq Y \leq y]} \geq \frac{P[y \leq X \leq v]}{P[y \leq Y \leq v]},$$

for all $x \leq u \leq y \leq v$ and, consequently, we see that

$$\frac{P[x \leq X \leq u]}{P[x \leq Y \leq u]} \geq \frac{P[y \leq X \leq v]}{P[y \leq Y \leq v]}, \qquad \text{for all } x \leq u \leq y \leq v.$$

The result follows taking limits as x tends to u and y to v in the previous inequality.

Let us prove now the inverse implication. Assume $X \leq_{lr} Y$. Then, it is easy to see that

$$[X|x \leq X \leq y] \leq_{lr} [Y|x \leq Y \leq y], \quad \text{for all } x \leq y,$$

and the result follows from (2.23). □

Observe that, taking limits as y tends to $+\infty$ in the thesis of the previous theorem, we have

$$X \leq_{lr} Y \Rightarrow X \leq_{hr} Y. \tag{2.25}$$

However, the inverse is not true in general, although it holds under some additional conditions. In particular, the following theorem is stated.

Theorem 2.5.3. *Let X and Y be two continuous random variables with survival functions $\overline{F}$ and $\overline{G}$ and hazard rate functions r and s, respectively. If $r(x)s(y) \geq r(y)s(x)$, for all $x \leq y$ such that $\overline{F}(x), \overline{F}(y), \overline{G}(x), \overline{G}(y) > 0$, and $X \leq_{hr} Y$, then*

$$X \leq_{lr} Y.$$

Proof. The proof follows easily from the equality

$$f(x)g(y) = r(x)\overline{F}(x)s(y)\overline{G}(y).$$ □

Repeatedly, the normal and the gamma distributions are examples of families with no explicit expression for the distribution functions. Let us compare two normal and gamma distributions in the likelihood ratio order.

Example 2.5.4. Let $X \sim N(\mu_1, \sigma_1)$ and $Y \sim N(\mu_2, \sigma_2)$. According to Example 2.4.16, we have $X \leq_{lr} Y$, whenever $\sigma_1 = \sigma_2$ and $\mu_1 \leq \mu_2$. Therefore, under these conditions and from (2.25), we also have that $X \leq_{hr} Y$.

Example 2.5.5. Let $X \sim G(\alpha_1, \beta_1)$ and $Y \sim G(\alpha_2, \beta_2)$ with density functions f and g, respectively. According to Example 2.4.17, we see that the ratio $g(x)/f(x)$ is increasing in $x \in (0, \infty)$, if $\alpha_1 \leq \alpha_2$ and $\beta_1 \leq \beta_2$. Therefore, we have $X \leq_{lr[hr]} Y$, whenever $\alpha_1 \leq \alpha_2$ and $\beta_1 \leq \beta_2$. Figure 2.13 shows a particular case of this situation.

To conclude this section, some preservation results for the likelihood ratio order are provided.

Theorem 2.5.6. *Let $\{X_n\}_{n \in \mathbb{N}}$ and $\{Y_n\}_{n \in \mathbb{N}}$ be two sequences of continuous random variables such that X_n converges in distribution to X and Y_n converges in distribution to Y. If $X_n \leq_{lr} Y_n$, for all $n \in \mathbb{N}$, then*

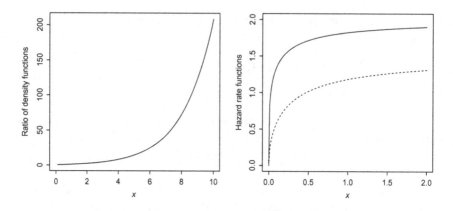

Figure 2.13 Ratio of the density functions of $X \sim G(1.25, 0.5)$ and $Y \sim G(1.5, 2/3)$ on the left side and hazard rate functions of $X \sim G(1.25, 0.5)$ (continuous line) and $Y \sim G(1.5, 2/3)$ (dashed line) on the right side.

$$X \leq_{lr} Y.$$

Proof. The proof follows under limiting arguments, taking into account the characterization given in Theorems 2.5.3 and 2.2.7. □

For a proof in the general case, the reader can look in Ref. [3].

Let us see now the preservation of the likelihood ratio under increasing transformations.

Theorem 2.5.7. *Let X and Y be two random variables. If $X \leq_{lr} Y$, then*

$$\phi(X) \leq_{lr} \phi(Y), \text{ for all real valued increasing function } \phi.$$

Proof. Let us assume that ϕ is strictly increasing. The result then follows from the equalities

$$P[x \leq \phi(X) \leq y] = P[\phi^{-1}(x) \leq X \leq \phi^{-1}(y)],$$

and

$$P[x \leq \phi(Y) \leq y] = P[\phi^{-1}(x) \leq Y \leq \phi^{-1}(y)],$$

and (2.24). If ϕ is just increasing, the result follows from Theorem 2.5.6 by limiting arguments as in previous sections. □

Next, a result on the preservation under mixtures of the likelihood ratio order is given.

Theorem 2.5.8. *Let $\{X(\theta), \theta \in S\} \subseteq \mathbb{R}$ be a family of random variables, and Θ_1 and Θ_2 be two random variables with common support S. If*

$$X(\theta) \leq_{lr} X(\theta'), \quad for \ all \ \theta \leq \theta' \in S, \tag{2.26}$$

and

$$\Theta_1 \leq_{lr} \Theta_2, \tag{2.27}$$

then,

$$X(\Theta_1) \leq_{lr} X(\Theta_2). \tag{2.28}$$

Proof. Let us denote by $f_\theta(x)$ the density function of $X(\theta)$ and by $h_i(\theta)$ the density function of Θ_i, for $i = 1, 2$. It is easy to see that $f_\theta(x)$ and $h_i(\theta)$ are TP_2 in (x, θ) and (θ, i), respectively, from conditions (2.26) and (2.27). It is also clear that the density function of $X(\Theta_i)$, for $i = 1, 2$, is given by

$$f_i(x) = \int_S f_\theta(x) h_i(\theta) \, d\theta.$$

Then, from Theorem 1.2.5, we find that $f_i(x)$ is TP_2 in (x, i) and this is equivalent to condition (2.28). □

Finally, some results on the preservation of the likelihood ratio order under convolutions are given.

Theorem 2.5.9. *Let X, Y and Z be three random variables such that Z is independent of X and Y. If $X \leq_{lr} Y$ and Z is ILR, then*

$$X + Z \leq_{lr} Y + Z.$$

Proof. The proof is similar to the one of Theorem 2.4.7 and follows as an application of Theorem 2.5.8. Let us consider the random variables $[X + Z|X = x]$ and $[Y + Z|Y = x]$, for all x, which are equally distributed from the independence assumption. It is not difficult to see that $[X + Z|X = x] \leq_{lr} [Y + Z|Y = x']$, for all $x \leq x'$, since Z is ILR. Therefore, if we consider the mixture of $[X + Z|X = x]$ and $[Y + Z|Y = x]$ with X and Y, we find that $X + Z \leq_{lr} Y + Z$. □

Repeated application of this theorem can be used to prove the following general result, in a similar way to Theorem 2.4.8, by the preservation under convolution of logconcave density functions (see Theorem 1.2.2).

Theorem 2.5.10. *Let $\{(X_i, Y_i)\}_{i=1}^n$ be independent pairs of continuous random variables. If X_i and Y_i are ILR and $X_i \leq_{lr} Y_i$, for all $i = 1, \ldots, n$, then*

$$\sum_{i=1}^{n} X_i \leq_{\text{lr}} \sum_{i=1}^{n} Y_i.$$

2.6 DISPERSIVE ORDERS

Another context where the stochastic orders arise is in the comparison of random variables in terms of their variability. The basic way to decide if one random variable has greater variability than another one, is comparing their variances. However, it is possible to provide more detailed methods to compare this characteristic by means of some stochastic orders. Probably, the most important one in this context is the dispersive order, which was introduced by Lewis and Thompson [64]. Let us see the definition of this criterion.

Definition 2.6.1. Given two random variables X and Y with quantile functions F^{-1} and G^{-1}, respectively, we say that X is smaller than Y in the *dispersive order*, denoted by $X \leq_{\text{disp}} Y$, if

$$G^{-1}(q) - G^{-1}(p) \geq F^{-1}(q) - F^{-1}(p), \quad \text{for all } 0 < p < q < 1. \quad (2.29)$$

Let us compare two Pareto distributions directly from the definition.

Example 2.6.2. Let $X \sim P(a_1, k_1)$ and $Y \sim P(a_2, k_2)$ with quantile functions F^{-1} and G^{-1}, respectively. Let us study the behavior of

$$G^{-1}(p) - F^{-1}(p) = \frac{k_2}{(1-p)^{1/a_2}} - \frac{k_1}{(1-p)^{1/a_1}},$$

which has a relative extreme at

$$p_0 = 1 - \left(\frac{k_2 a_1}{k_1 a_2}\right)^{\frac{a_1 a_2}{a_1 - a_2}}.$$

It is easy to see that $p_0 \notin (0, 1)$, if $a_1 \geq a_2$ and $a_1 k_2 \geq a_2 k_1$. Therefore, if $a_1 \geq a_2$ and $a_1 k_2 \geq a_2 k_1$, then $X \leq_{\text{disp}} Y$. Figure 2.14 shows a particular example of this situation.

The definition of this criterion leads to an immediate interpretation. The dispersive order compares the distance between any pair of quantiles of X and Y, that means, if $X \leq_{\text{disp}} Y$, the quantiles of X are less spread than

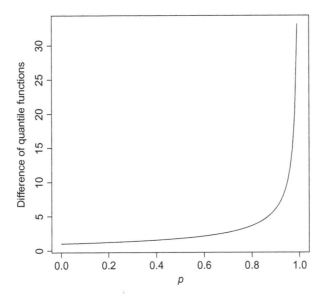

Figure 2.14 Difference of the quantile functions of $X \sim P(2, 1)$ and $Y \sim P(1.5, 2)$.

the corresponding ones of Y. Let us observe that (2.29) is equivalent to the condition

$$G^{-1}(p) - F^{-1}(p) \text{ be increasing in } p \in (0, 1).$$

If X and Y are continuous, replacing $p = F(x)$ in (2.29) we find, equivalently, that the dispersive order holds if, and only if,

$$G^{-1}(F(x)) - x \text{ is increasing in } x \text{ in the support of } X. \tag{2.30}$$

Moreover, if X and Y have open interval supports with differentiable distributions and density functions f and g, respectively, the previous condition can be equivalently written as

$$f(F^{-1}(p)) \geq g(G^{-1}(p)), \quad \text{for all } p \in (0, 1), \tag{2.31}$$

or, equivalently,

$$r(F^{-1}(p)) \geq s(G^{-1}(p)), \quad \text{for all } p \in (0, 1),$$

where r and s denote the hazard rate functions of X and Y, respectively.

Notice that the dispersive order is location free, that means, if $X \leq_{\text{disp}} Y$, then $X + c \leq_{\text{disp}} Y$ and $X \leq_{\text{disp}} Y + c$, for all $c \in \mathbb{R}$.

The following characterization, based on dispersive functions, reinforces the use of the dispersive order as a tool to compare variability. Recall first that a real valued function ϕ is dispersive if $\phi(y) - \phi(x) \geq y - x$, for all $x \leq y$. Observe that a dispersive function is always strictly increasing.

Theorem 2.6.3. *Let X and Y be two continuous random variables. Then, $X \leq_{\text{disp}} Y$ if, and only if, a dispersive function ϕ exists such that $Y =_{\text{st}} \phi(X)$.*

Proof. Let us consider the function $\phi(x) = G^{-1}(F(x))$. From (2.30), we see that ϕ is a dispersive function and, from the property (ii) of the quantile function, $Y =_{\text{st}} \phi(X)$. Let us assume now that $Y =_{\text{st}} \phi(X)$, where ϕ is a dispersive function. Then, from the property v) of the quantile function, we see that $G^{-1}(p) = \phi(F^{-1}(p))$ and the proof follows from the dispersive property of ϕ. □

Another important characterization of the dispersive order is the following [65].

Theorem 2.6.4. *Let X and Y be two random variables, with quantile functions F^{-1} and G^{-1}, respectively. Then $X \leq_{\text{disp}} Y$ if, and only if,*

$$(X - F^{-1}(p))_+ \leq_{\text{st}} (Y - G^{-1}(p))_+, \quad \text{for all } p \in (0, 1). \quad (2.32)$$

Proof. First, observe that the distribution function of $(X - F^{-1}(p))_+$ is given by

$$F_p^+(x) = \begin{cases} 0 & \text{if } x < 0, \\ F(x + F^{-1}(p)) & \text{if } x \geq 0, \end{cases} \quad (2.33)$$

and analogously for $(Y - G^{-1}(p))_+$. Therefore, (2.32) is equivalent to the condition

$$F(x + F^{-1}(p)) \geq G(x + G^{-1}(p)), \quad \text{for all } x \geq 0 \text{ and } p \in (0, 1). \quad (2.34)$$

Let us assume first $X \leq_{\text{disp}} Y$. We shall show that (2.34) holds. Let us consider fixed values $x \geq 0$, $p \in (0, 1)$ and $q = G(x + G^{-1}(p))$. Then, we have the following chain of inequalities:

$$\begin{aligned} G(x + G^{-1}(p)) &= q \leq F(F^{-1}(q)) \\ &= F([F^{-1}(q) - F^{-1}(p)] + F^{-1}(p)) \\ &\leq F([G^{-1}(q) - G^{-1}(p)] + F^{-1}(p)) \\ &\leq F(x + F^{-1}(p)), \end{aligned}$$

where the first inequality follows from the property (iii) of the quantile function, the second one follows from the definition of the dispersive order and the last one follows from the property (iv) of the quantile function.

Let us assume now that (2.34) holds and consider $0 < p < q < 1$. Once more, we have the following chain of inequalities:

$$q \leq G(G^{-1}(q)) = G\left(\left[G^{-1}(q) - G^{-1}(p)\right] + G^{-1}(p)\right)$$
$$\leq F\left(\left[G^{-1}(q) - G^{-1}(p)\right] + F^{-1}(p)\right),$$

where the last inequality follows replacing $x = G^{-1}(q) - G^{-1}(p)$ in (2.34). Therefore, $F^{-1}(q) \leq G^{-1}(q) - G^{-1}(p) + F^{-1}(p)$, for all $0 < p < q < 1$ and, consequently, $X \leq_{\mathrm{disp}} Y$. $\qquad\square$

Recalling the interpretation of the random variable $(X - F^{-1}(p))_+$ seen in Section 1.2, this result provides a useful interpretation of the dispersive order in both risk theory and reliability.

Whenever F and G are strictly increasing in $F^{-1}(p)$ and $G^{-1}(p)$, respectively, it is also possible to prove that condition (2.34) just requires to be verified for all $p \in (0, 1)$ and for all $x > 0$ such that $F^{-1}(p) + x \in C(F)$ and $G^{-1}(p) + x \in C(G)$. The proof is rather technical and it can be seen in Ref. [64].

From the definition of the dispersive order and Theorem 2.2.3, it is easy to provide the following relationships between the dispersive and the stochastic orders.

Theorem 2.6.5. *Let X and Y be two random variables with quantile functions F^{-1} and G^{-1}, respectively, such that $X \leq_{\mathrm{disp}} Y$.*

(i) *If $\lim_{p \to 0^+}(G^{-1}(p) - F^{-1}(p)) \geq 0$, then $X \leq_{\mathrm{st}} Y$.*
(ii) *If $\lim_{p \to 1^-}(G^{-1}(p) - F^{-1}(p)) \leq 1$, then $Y \leq_{\mathrm{st}} X$.*

Notice that the previous conditions on the limits can be rewritten in terms of the left and the right extremes of the supports, respectively, whenever they are finite.

Next, the relationship between the dispersive and the hazard rate orders is established [66, 67].

Theorem 2.6.6. *Let X and Y be two continuous random variables with distribution functions F and G, respectively.*

(i) *If $X \leq_{\mathrm{hr}} Y$ and X or Y or both are DFR, then $X \leq_{\mathrm{disp}} Y$.*
(ii) *If $X \leq_{\mathrm{disp}} Y$, $\lim_{p \to 0^+}(G^{-1}(p) - F^{-1}(p)) \geq 0$ and X or Y or both are IFR, then $X \leq_{\mathrm{hr}} Y$.*

Proof. First, we prove (i). Let us consider that X is DFR, if Y is DFR the proof is similar. Recall that the hazard rate order implies the stochastic order and, therefore, it holds that $F^{-1}(G(t)) \leq t$, for all t. Hence,

$$\frac{\overline{F}(F^{-1}(G(t)) + x)}{\overline{F}(F^{-1}(G(t)))} \leq \frac{\overline{F}(t+x)}{\overline{F}(t)}, \quad \text{for all } t \text{ and } x > 0,$$

by the DFR property. Furthermore, we have

$$\frac{\overline{F}(t+x)}{\overline{F}(t)} \leq \frac{\overline{G}(t+x)}{\overline{G}(t)}, \quad \text{for all } t \text{ and } x > 0,$$

by Theorem 2.4.3. Therefore, we get

$$\frac{\overline{F}(F^{-1}(G(t)) + x)}{\overline{F}(F^{-1}(G(t)))} \leq \frac{\overline{G}(t+x)}{\overline{G}(t)}, \quad \text{for all } t \text{ and } x > 0.$$

Since $F(F^{-1}(G(t))) \geq G(t)$ by the property (iii) of the quantile function, then $F(F^{-1}(G(t)) + x) \geq G(t+x)$, for all t and $x > 0$. Replacing $G(t) = p$, we have $F(F^{-1}(p) + x) \geq G(G^{-1}(p) + x)$, for all $p \in (0,1)$ and $x > 0$, which is equivalent to $X \leq_{\text{disp}} Y$ by (2.34).

Now, we prove (ii). Let us consider that X is IFR, if Y is IFR the proof is similar. Analogously to (2.30), it is easy to see that $X \leq_{\text{disp}} Y$ holds if, and only if, $F^{-1}(G(x)) - x$ is decreasing in x. Therefore, we have $F^{-1}(G(x+t)) \leq F^{-1}(G(t)) + x$, which implies $G(x+t) \leq F(F^{-1}(G(t)) + x)$. Combining this inequality with the property (iii) of the quantile function, we see that

$$\frac{\overline{F}(F^{-1}(G(t)) + x)}{\overline{F}(F^{-1}(G(t)))} \leq \frac{\overline{G}(x+t)}{\overline{G}(t)}. \tag{2.35}$$

Moreover, combining $F^{-1}(G(t)) \leq t$ (which holds from Theorem 2.6.5) with the IFR property, we see that

$$\frac{\overline{F}(t+x)}{\overline{F}(t)} \leq \frac{\overline{F}(F^{-1}(G(t)) + x)}{\overline{F}(F^{-1}(G(t)))}. \tag{2.36}$$

From (2.35) and (2.36), we get $X \leq_{\text{hr}} Y$, by Theorem 2.4.3. □

Next, we give a characterization of the dispersive order in terms of the sign changes of certain set of functions [68, 69].

Theorem 2.6.7. *Let X and Y be two continuous random variables with distribution functions F and G, respectively. Then, $X \leq_{\text{disp}} Y$ if, and only if,*

$$S^{-}(G(\cdot) - F(\cdot - c)) \leq 1 \text{ with the sign sequence } +, - \text{ in case of equality,}$$

for all $c \in \mathbb{R}$.

Proof. Assume first that $X \leq_{disp} Y$. Let us suppose that some c exists such that $F(x_0 - c) = G(x_0)$ and

$$F(x - c) < G(x), \quad \text{for all } x \in (x_0, x_0 + \epsilon], \epsilon > 0.$$

If we denote by $p = F(x_0 - c)$ and $q = F(x_0 - c + \epsilon)$, then $G(x_0 + \epsilon) > q > p = G(x_0)$ and, consequently, $G^{-1}(q) - G^{-1}(p) < F^{-1}(q) - F^{-1}(p)$, which contradicts $X \leq_{disp} Y$.

Assume now that $S^{-}(G(\cdot) - F(\cdot - c)) \leq 1$ with the sign sequence $+, -$ in case of equality, for all $c \in \mathbb{R}$. First, we shall prove that this condition implies either $X \leq_{disp} Y$ or $X \geq_{disp} Y$. Let us suppose the opposite—that is, $p_i < q_i \in (0, 1)$ exist, for $i = 1, 2$, such that

$$\begin{cases} F^{-1}(q_1) - F^{-1}(p_1) > G^{-1}(q_1) - G^{-1}(p_1), \\ F^{-1}(q_2) - F^{-1}(p_2) < G^{-1}(q_2) - G^{-1}(p_2). \end{cases} \tag{2.37}$$

Therefore, the sequence $\{F^{-1}(r_i) - G^{-1}(r_i)\}_{i=1}^{4}$ is neither increasing nor decreasing, where r_i, for $i = 1, 2, 3, 4$ denote the ordered values p_i, q_i, for $i = 1, 2$. Let us suppose without lost of generality that

$$F^{-1}(r_1) - G^{-1}(r_1) \geq F^{-1}(r_2) - G^{-1}(r_2) \leq F^{-1}(r_3) - G^{-1}(r_3),$$

since, at least, three of the $r_i's$ are different, otherwise it is not possible that (2.37) holds. Let us consider now

$$c = -\frac{1}{2} \min \left\{ F^{-1}(r_1) - G^{-1}(r_1), F^{-1}(r_3) - G^{-1}(r_3) + F^{-1}(r_2) - G^{-1}(r_2) \right\}.$$

Taking into account that the quantile function of $X + c$ is given by $F_c^{-1}(p) = F^{-1}(p) + c$, we find that

$$F_c^{-1}(r_1) > G^{-1}(r_1), F_c^{-1}(r_2) < G^{-1}(r_2) \text{ and } F_c^{-1}(r_3) > G^{-1}(r_3).$$

Taking $x_1 = G^{-1}(r_1), x_2 = G^{-1}(r_2)$ and $x_3 = G^{-1}(r_3)$, we see that

$$F_c(x_1) < G(x_1), F_c(x_2) > G(x_2) \text{ and } F_c(x_3) < G(x_3),$$

which means $S^{-}(G(\cdot) - F(\cdot - c)) \geq 2$, which contradicts the hypothesis.

Next, we show the sense the dispersive order holds in. On the one hand, if $S^{-}(G(\cdot) - F(\cdot - c)) = 0$, for all $c \in \mathbb{R}$, then it is easy to see that a real value c exists such that $F_c = G$ and, consequently, both $X \leq_{disp} Y$ and $X \geq_{disp} Y$ hold. On the other hand, if c exists such that $S^{-}(G(\cdot) - F(\cdot - c)) = 1$ with the sign sequence $+, -$, then $x < y \in \mathbb{R}$ exist such that $F_c(x) < G(x)$ and $F_c(y) > G(y)$. Now, let denote by $p = G(x)$ and $q = F_c(y)$, we see that

$$G^{-1}(q) - G^{-1}(p) > F_c^{-1}(q) - F_c^{-1}(p) = F^{-1}(q) - F^{-1}(p),$$

hence $X \leq_{\text{disp}} Y$.   □

If the random variables have their support in $[0, +\infty)$, it is possible to weaken the condition on the sign changes, as is stated by the upcoming theorem. The proof follows easily from the previous theorem and Theorem 2.6.5.

Theorem 2.6.8. *Let X and Y be two continuous random variables with distribution functions F and G, respectively and common support $(0, +\infty)$. Then, $X \leq_{\text{disp}} Y$ if, and only if,*

(i) $X \leq_{\text{st}} Y$, and
(ii) $S^-(G(\cdot) - F(\cdot - c)) \leq 1$ with the sign sequence $+, -$ in case of equality, for all $c > 0$.

As for the previous criteria, sets of sufficient conditions exist for the dispersive order in terms of the density functions. The proof of the upcoming theorems can be easily obtained from the previous ones and Corollary 1.2.4.

Theorem 2.6.9. *Let X and Y be two continuous random variables with density functions f and g, respectively. If $S^-(g(\cdot) - f(\cdot - c)) \leq 2$ with the sign sequence $+, -, +$ in case of equality, for all $c \in \mathbb{R}$, then*

$$X \leq_{\text{disp}} Y.$$

Theorem 2.6.10. *Let X and Y be two continuous random variables with density functions f and g, respectively, and common support $(0, +\infty)$. If $S^-(g(\cdot) - f(\cdot - c)) \leq 2$ with the sign sequence $+, -, +$ in case of equality, for all $c > 0$, and $X \leq_{\text{st}} Y$, then*

$$X \leq_{\text{disp}} Y.$$

The previous result will be used to provide examples of parametric families ordered in the dispersive order. Previously, some additional results on the preservation of the dispersive order under transformations are required. In fact, the remaining theorems are preservation results. First, the result on preservation under weak convergence is introduced.

Theorem 2.6.11. *Let $\{X_n\}_{n\in\mathbb{N}}$ and $\{Y_n\}_{n\in\mathbb{N}}$ be two sequences of random variables such that X_n converges in distribution to X and Y_n converges in distribution to Y. If $X_n \leq_{disp} Y_n$, for all $n \in \mathbb{N}$, then*

$$X \leq_{disp} Y.$$

Proof. First, we recall that, if a sequence of distribution functions $\{F_n\}_{n\in\mathbb{N}}$ converges weakly to a distribution function F, then the sequence $\{F_n^{-1}\}_{n\in\mathbb{N}}$ converges weakly to F^{-1} [64]. To show that $X \leq_{disp} Y$, we shall prove the equivalent condition (2.34). From previous comments, we just need to verify (2.34), for all $p \in (0,1)$ such that F and G are strictly increasing in $F^{-1}(p)$ and $G^{-1}(p)$, respectively, and for all $x > 0$ such that $F^{-1}(p) + x \in C(F)$ and $G^{-1}(p) + x \in C(G)$. Let us consider $p \in (0,1)$ and $x > 0$ satisfying these conditions and let us denote by F_n and G_n the distribution function of X_n and Y_n, respectively, for all $n \in \mathbb{N}$. From the hypothesis, we see that

$$F_n(x + F_n^{-1}(p)) \geq G_n(x + G_n^{-1}(p)).$$

The proof follows observing that $F_n(x + F_n^{-1}(p)) \to F(x + F^{-1}(p))$ and $G_n(x + G_n^{-1}(p)) \to G(x + G^{-1}(p))$. $\square$

Let us see now some preservation results under transformations.

Theorem 2.6.12. *Let X and Y be two continuous random variables with differentiable distribution functions, such that $X \leq_{st} Y$. If $X \leq_{disp} Y$, then*

$$\phi(X) \leq_{disp} \phi(Y), \text{ for all real valued increasing convex function } \phi.$$

Proof. Let us denote by F [G] and f [g] the distribution and density functions of X [Y], respectively. Besides, let us denote by $F_{\phi(X)}^{-1}$ [$G_{\phi(Y)}^{-1}$] and $f_{\phi(X)}$ [$g_{\phi(Y)}$] the quantile and density functions of $\phi(X)$ [$\phi(Y)$], respectively. Let us assume that ϕ is differentiable (otherwise ϕ could be obtained by approximation of differentiable functions and the result would follow from the previous theorem). Notice that

$$f_{\phi(X)}(F_{\phi(X)}^{-1}(p)) = f(F^{-1}(p))/\phi'(F^{-1}(p)),$$

and analogously for $\phi(Y)$. Let us observe that ϕ' is strictly positive, since ϕ is strictly increasing. On the one hand, we have $f(F^{-1}(p)) \geq g(G^{-1}(p))$, by (2.31). On the other hand, we see that $\phi'(F^{-1}(p)) \leq \phi'(G^{-1}(p))$, since $X \leq_{st} Y$ and ϕ' is increasing. Therefore,

$$f_{\phi(X)}(F_{\phi(X)}^{-1}(p)) = \frac{f(F^{-1}(p))}{\phi'(F^{-1}(p))} \geq \frac{g(G^{-1}(p))}{\phi'(G^{-1}(p))} = g_{\phi(X)}(G_{\phi(Y)}^{-1}(p)),$$

for all $p \in (0,1)$ and, consequently, $\phi(X) \leq_{\text{disp}} \phi(Y)$, again by (2.31). □

Theorem 2.6.13. *Let X be a random variable and ϕ_1 and ϕ_2 be two strictly increasing functions. If $\phi_1(y) - \phi_1(x) \leq \phi_2(y) - \phi_2(x)$, for all $x \leq y$, then $\phi_1(X) \leq_{\text{disp}} \phi_2(X)$.*

Proof. First, we observe that the quantile function of $\phi_i(X)$ is given by $\phi_i(F^{-1}(p))$, for $i = 1, 2$ and for all $p \in (0,1)$, where F^{-1} is the quantile function of X. Then, the result follows by the properties of ϕ_1 and ϕ_2. □

Next, we combine the sufficient conditions given by Theorems 2.6.9 and 2.6.10 with the results on the preservation under transformations of the dispersive order, to compare two gamma and two normal distributions.

Example 2.6.14. Let $X \sim G(\alpha_1, \beta_1)$ and $Y \sim G(\alpha_2, \beta_2)$ with density functions f and g, respectively. First, we assume $\beta_1 = \beta_2 = 1$ and denote $X_{\alpha_1} \sim G(\alpha_1, 1)$ and $X_{\alpha_2} \sim G(\alpha_2, 1)$. It is easy to see that $S^-(g(\cdot) - f(\cdot - c)) = S^-(h_c(x))$, where

$$h_c(x) = \log\left(\frac{\Gamma(\alpha_1)}{\Gamma(\alpha_2)}\right) - (\alpha_1 - 1)\log(x - c) - c + (\alpha_2 - 1)\log(x).$$

If $\alpha_1 < \alpha_2$, the function $h_c(x)$ attains a relative extreme at

$$x_0 = \frac{c(1 - \alpha_2)}{\alpha_1 - \alpha_2},$$

hence $S^-(h_c(x)) \leq 2$. Furthermore, it is not difficult to see that $\lim_{x \to +\infty} h_c(x) \geq 0$ and, therefore, we get the sign sequence $+, -, +$, in case of equality. If $\alpha_1 = \alpha_2$, then $h_c(x)$ is monotone and, consequently, $S^-(h_c(x)) \leq 1$. Therefore, from Example 2.5.5, (2.23), and Theorem 2.6.10, we can conclude $X_{\alpha_1} \leq_{\text{disp}} X_{\alpha_2}$. Now, let us consider the transformation $\phi_1(x) = \beta_1 x$, then $X =_{\text{st}} \phi_1(X_{\alpha_1}) \leq_{\text{disp}} \phi_1(X_{\alpha_2})$, by Theorem 2.6.12. Next, we take $\phi_2(x) = \beta_2 x$ and assume $\beta_1 \leq \beta_2$, then $\phi_1(X_{\alpha_2}) \leq_{\text{disp}} \phi_2(X_{\alpha_2}) =_{\text{st}} Y$, by Theorem 2.6.13. In conclusion, we have $X \leq_{\text{disp}} Y$, whenever $\alpha_1 \leq \alpha_2$ and $\beta_1 \leq \beta_2$. Figure 2.15 shows a particular case of this situation.

Example 2.6.15. Let $X \sim N(\mu_1, \sigma_1)$ and $Y \sim N(\mu_2, \sigma_2)$. Let us consider the random variable $Z \sim N(0,1)$ and the transformations $\phi_i(x) = \sigma_i x + \mu_i$, for $i = 1, 2$, then, $X = \phi_1(Z)$ and $Y = \phi_2(Z)$. If $\sigma_1 \leq \sigma_2$, it is easy to see

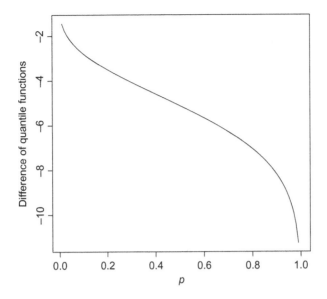

Figure 2.15 Difference of the quantile functions of $X \sim G(2, 4/3)$ and $Y \sim G(4, 2)$.

that the condition $\phi_1(y) - \phi_1(x) \le \phi_2(y) - \phi_2(x)$, for all $x \le y$, is verified and, therefore, according to Theorem 2.6.13, we get $X \le_{\mathrm{disp}} Y$, whenever $\sigma_1 \le \sigma_2$.

Finally, a result on the preservation of the dispersive order under convolutions is given.

Theorem 2.6.16. *Let $X_1, \ldots, X_n$ and $Y_1, \ldots, Y_n$ be two sets of independent random variables such that X_i and Y_i are ILR and $X_i \le_{\mathrm{disp}} Y_i$, for all $i = 1, \ldots, n$, then*

$$\sum_{i=1}^{n} X_i \le_{\mathrm{disp}} \sum_{i=1}^{n} Y_i.$$

Proof. The proof is based on the fact that X is ILR if, and only if, $X + Z \le_{\mathrm{disp}} X + Y$, for any pair of random variables X, Y independent of Z [64]. Now, by iteration of this result and the preservation under convolution of logconcave density functions (see Theorem 1.2.2), we get the result. □

Similarly to the definition of the mean residual life order through weakening the condition of the hazard rate order, a weaker criterion to compare the variability of two random variables can be defined. By (2.2), if we consider the comparison of the means of the random variables

$(X - F^{-1}(p))_+$ and $(Y - G^{-1}(p))_+$, for all $p \in (0, 1)$, a weaker criterion arises. Since the expected values of these random variables define the excess wealth functions (see Section 1.2), this condition leads to the so-called excess wealth order.

Definition 2.6.17. Given two random variables X and Y, we say that X is smaller than Y in the *excess wealth order*, denoted by $X \leq_{ew} Y$, if

$$W_X(p) \leq W_Y(p), \quad \text{for all } p \in (0, 1).$$

The excess wealth order is also known as the *right spread order* [70]. From the relationship between the excess wealth function and the CVaR measure, we see that $X \leq_{ew} Y$ is equivalent to

$$\text{CVaR}\,[X;p] \leq \text{CVaR}\,[Y;p], \quad \text{for all } p \in (0, 1),$$

for all continuous random variables. Therefore, from the interpretation of these measures in risk theory, given in Section 1.2, the excess wealth order has a clear interpretation in this context. Furthermore, recalling that $(X - F^{-1}(p))_+$ can be related to burn-in techniques in reliability theory, the dispersive and the excess wealth orders can be also interpreted in this context.

From (1.6), it holds $X \leq_{ew} Y$ if, and only if,

$$\frac{1}{1-p} \int_p^1 (G^{-1}(u) - F^{-1}(u))\, du \geq G^{-1}(p) - F^{-1}(p), \quad \text{for all } p \in (0, 1).$$

As pointed out, from the definition of the dispersive order and the previous characterization, it is clear that

$$X \leq_{disp} Y \Rightarrow X \leq_{ew} Y. \tag{2.38}$$

The following characterization of the excess wealth order shows that this criterion does not only compare the expected values of $(X - F^{-1}(p))_+$ and $(Y - G^{-1}(p))_+$, for all $p \in (0, 1)$, but also the expected values of any real valued increasing convex transformation of these random variables.

Theorem 2.6.18. *Let X and Y be two continuous random variables with quantile functions F^{-1} and G^{-1}, respectively. Then, $X \leq_{ew} Y$ if, and only if,*

$$\left(X - F^{-1}(p)\right)_+ \leq_{icx} \left(Y - G^{-1}(p)\right)_+, \quad \text{for all } p \in (0, 1).$$

Proof. Since the increasing convex order implies that the means are ordered, the inverse implication is trivial. Let us assume now that $X \leq_{\text{ew}} Y$. The condition $\left(X - F^{-1}(p)\right)_+ \leq_{\text{icx}} \left(Y - G^{-1}(p)\right)_+$, for all $p \in (0,1)$ is equivalent to

$$\int_{x+F^{-1}(p)}^{+\infty} \overline{F}(t)dt \leq \int_{x+G^{-1}(p)}^{+\infty} \overline{G}(t)dt, \quad \text{for all } x \geq 0 \text{ and } p \in (0,1),$$

(2.39)

by (2.33) and (2.5). If we define the function

$$H(x,p) = \int_{x+G^{-1}(p)}^{+\infty} \overline{G}(t)\, dt - \int_{x+F^{-1}(p)}^{+\infty} \overline{F}(t)\, dt,$$

for all $x \geq 0$ and $p \in (0,1)$, then (2.39) is equivalent to prove that $H(x,p) \geq 0$, for all $x \geq 0$ and $p \in (0,1)$. First, we observe that

$$\lim_{x \to +\infty} H(x,p) = 0, \quad \text{for all } p \in (0,1).$$

(2.40)

Hence, it is enough to show that $H(x_0,p) \geq 0$, for a fixed $p \in (0,1)$ where x_0 denotes a relative minimum of H, or equivalently, $\left.\frac{\partial H(x,p)}{\partial x}\right|_{x=x_0} = 0$, since $H(x,p)$ is continuous and differentiable. The previous equality holds if, and only if,

$$F(x_0 + F^{-1}(p)) = G(x_0 + G^{-1}(p)) = q.$$

Assuming F and G are strictly increasing in $x_0 + F^{-1}(p)$ and $x_0 + G^{-1}(p)$, respectively (otherwise, the proof is more technical and it is omitted [71]), we see that $F^{-1}(q) = x_0 + F^{-1}(p)$ and $G^{-1}(q) = x_0 + G^{-1}(p)$. Therefore,

$$H(x_0,p) = \int_{x_0+G^{-1}(p)}^{\infty} \overline{G}(t)\, dt - \int_{x_0+F^{-1}(p)}^{\infty} \overline{F}(t)\, dt$$

$$= \int_{G^{-1}(q)}^{\infty} \overline{G}(t)\, dt - \int_{F^{-1}(q)}^{\infty} \overline{F}(t)\, dt \geq 0,$$

due to the fact that $X \leq_{\text{ew}} Y$.   □

If X and Y have finite left end points of their supports such that $l_X \leq l_Y$, taking limits as p tends to 0^+ in the previous theorem, we see that

$$X \leq_{\text{ew}} Y \Rightarrow X \leq_{\text{icx}} Y.$$

Under additional assumptions, the excess wealth order is also related to the mrl order. Next, the result is stated without proof (see Ref. [72] for details).

Theorem 2.6.19. *Let X and Y be two continuous random variables with quantile functions F^{-1} and G^{-1}, respectively, and finite left extreme points of their supports such that $l_X \leq l_Y$.*

(i) If $X \leq_{mrl} Y$ and X or Y or both are IMRL, then $X \leq_{ew} Y$.
(ii) If $X \leq_{ew} Y$ and X or Y or both are DMRL, then $X \leq_{mrl} Y$.

One of the main problems for the applicability of the excess wealth order is the evaluation of incomplete integrals of quantile functions, which is not possible in most cases. Next, several results providing sufficient conditions for the excess wealth order, when the dispersive order does not hold, are given as well as examples where they are applied. These results can be found in Ref. [73]. Let us see the first set of sufficient conditions.

Theorem 2.6.20. *Let X and Y be two random variables with quantile functions F^{-1} and G^{-1}, respectively, and finite means. If a value $p_0 \in (0, 1)$ exists such that $G^{-1}(p) - F^{-1}(p) \leq E[Y] - E[X]$, for all $p \in (0, p_0)$ and $G^{-1}(p) - F^{-1}(p)$ is increasing in $p \in [p_0, 1)$, then*

$$X \leq_{ew} Y.$$

Proof. Let us consider a value $p \in [p_0, 1)$. It is easy to see that the quantile function of $(X - F^{-1}(p))_+$ is given by

$$F_+^{-1}(q, p) = \left(F^{-1}(q) - F^{-1}(p)\right)_+, \quad \text{for all } q \in (0, 1),$$

and similarly for the quantile function of $(Y - G^{-1}(p))_+$. Since $G^{-1}(p) - F^{-1}(p)$ is increasing in $p \in [p_0, 1)$, from Theorem 2.2.3, we have

$$(X - F^{-1}(p))_+ \leq_{st} (Y - G^{-1}(p))_+, \quad \text{for all } p \in [p_0, 1),$$

and, therefore, by (2.2),

$$E[(X - F^{-1}(p))_+] \leq E[(Y - G^{-1}(p))_+], \quad \text{for all } p \in [p_0, 1). \quad (2.41)$$

Let us consider now a value $p \in (0, p_0)$. We observe that the quantile function of $\min\{X, F^{-1}(p)\} - E[X]$ is given by

$$F_p^{-1}(q) = \begin{cases} F^{-1}(q) - E[X] & \text{if } 0 < q < p, \\ F^{-1}(p) - E[X] & \text{if } p \leq q < 1, \end{cases}$$

and similarly for $\min\{Y, G^{-1}(p)\} - E[Y]$. Since $G^{-1}(p) - F^{-1}(p) \leq E[Y] - E[X]$, for all $p \in (0, p_0)$, from Theorem 2.2.3, we have

$$\min\{X, F^{-1}(p)\} - E[X] \geq_{st} \min\{Y, G^{-1}(p)\} - E[Y],$$

for all $p \in (0, p_0)$, and, consequently, by (2.2),

$$E[\min\{X, F^{-1}(p)\}] - E[X] \geq E[\min\{Y, G^{-1}(p)\}] - E[Y],$$

for all $p \in (0, p_0)$. From the equality $(x - t)_+ = x - \min\{x, t\}$, we conclude

$$E[(X - F^{-1}(p))_+] \leq E[(Y - G^{-1}(p))_+], \quad \text{for all } p \in (0, p_0). \quad (2.42)$$

The result follows from (2.41) and (2.42). □

Notice that the assumption $G^{-1}(p) - F^{-1}(p) \leq E[Y] - E[X]$, for all $p \in (0, p_0)$ is trivially satisfied if a point p_0 exists such that $G^{-1}(p) \leq F^{-1}(p)$, for all $p \in (0, p_0)$ and $E[X] \leq E[Y]$.

Let us see an example where the previous result is applied to compare two Davies distributions.

Example 2.6.21. Let $X \sim D(\lambda_1, \theta_1, C_1)$ and $Y \sim D(\lambda_2, \theta_2, C_2)$ with quantile functions F^{-1} and G^{-1}, respectively. In order to have finite means, we assume $\theta_1, \theta_2 < 1$. To simplify the calculus, we study the monotonicity of $G^{-1}(p)/F^{-1}(p)$, or equivalently, the monotonicity of

$$H(p) = (\lambda_2 - \lambda_1) \log(p) - (\theta_2 - \theta_1) \log(1 - p).$$

It is easy to see that $H(p)$ is increasing in $p \in (0, 1)$, $\lim_{p \to 0^+} G^{-1}(p)/F^{-1}(p) = 0$ and $\lim_{p \to 1^-} G^{-1}(p)/F^{-1}(p) = +\infty$, whenever $\lambda_1 \leq \lambda_2$ and $\theta_1 \leq \theta_2$. Therefore, a value p_0 exists such that $G^{-1}(p) \leq F^{-1}(p)$, for all $p \in (0, p_0)$. In addition, since $G^{-1}(p) - F^{-1}(p) = F^{-1}(p)(G^{-1}(p)/F^{-1}(p) - 1)$, we see that $G^{-1}(p) - F^{-1}(p)$ is increasing in $p \in [p_0, 1)$. To sum up, if $\lambda_1 \leq \lambda_2$, $\theta_1 \leq \theta_2$ and

$$E[X] = C_1 B(1 + \lambda_1, 1 - \theta_1) \leq E[Y] = C_2 B(1 + \lambda_2, 1 - \theta_2),$$

we have $X \leq_{\text{ew}} Y$, but $X \not\leq_{\text{disp}} Y$ or $X \not\geq_{\text{disp}} Y$, since it is also possible to see that, under the previous conditions on the parameters, $G^{-1}(p) - F^{-1}(p)$ is decreasing in $p \in (0, p_0)$.

The following corollary can be easily obtained as a direct consequence of Theorem 2.6.20.

Corollary 2.6.22. *Let X and Y be two random variables with finite means such that* $\lim_{p \to 0^+} (G^{-1}(p) - F^{-1}(p)) \leq E[Y] - E[X]$. *Let us assume that a value* $p_0 \in (0, 1)$ *exists such that* $G^{-1}(p) - F^{-1}(p)$ *is decreasing in* $p \in (0, p_0)$ *and* $G^{-1}(p) - F^{-1}(p)$ *is increasing in* $p \in [p_0, 1)$, *then*

$$X \leq_{\text{ew}} Y.$$

This is the case of two Pareto and Weibull distributed random variables, as we see next.

Example 2.6.23. Let $X \sim P(a_1, k_1)$ and $Y \sim P(a_2, k_2)$ with quantile functions F^{-1} and G^{-1}, respectively. According to Example 2.6.2, if $a_1 > a_2$ and $a_1 k_2 > a_2 k_1$, then $G^{-1}(p) - F^{-1}(p)$ has a minimum. Therefore, if

$$\lim_{p \to 0^+} (G^{-1}(p) - F^{-1}(p)) = k_2 - k_1 \leq \frac{a_2 k_2}{a_2 - 1} - \frac{a_1 k_1}{a_1 - 1} = E[Y] - E[X],$$

$a_1 > a_2$ and $a_1 k_2 \geq a_2 k_1$, then $X \leq_{ew} Y$ but $X \not\leq_{disp} Y$ or $X \not\geq_{disp} Y$. Figure 2.16 shows a particular case of this situation.

Example 2.6.24. Let $X \sim W(\alpha_1, \beta_1)$ and $Y \sim W(\alpha_2, \beta_2)$ with quantile functions F^{-1} and G^{-1}, respectively. According to (2.30), we study equivalently the monotonicity of

$$H(x) = G^{-1}(F(x)) - x = \left(\frac{\alpha_2}{\alpha_1}\right)^{\frac{1}{\beta_2}} x^{\frac{\beta_1}{\beta_2}} - x, \quad \text{for all } x \geq 0.$$

By differentiation, we see that $H(x)$ attains its minimum at

$$x_0 = \left(\frac{\beta_2}{\beta_1} \left(\frac{\alpha_1}{\alpha_2}\right)^{\frac{1}{\beta_2}}\right)^{\frac{\beta_2}{\beta_1 - \beta_2}},$$

whenever $\beta_1 > \beta_2$. Furthermore, it can be seen easily that $\lim_{p \to 0^+}(G^{-1}(p) - F^{-1}(p)) = 0$ and $\lim_{p \to 1^-}(G^{-1}(p) - F^{-1}(p)) = +\infty$. Therefore, if we assume that $\beta_1 > \beta_2$ and

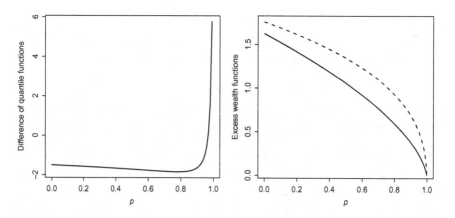

Figure 2.16 *Difference of the quantile functions of $X \sim P(2.5, 2.5)$ and $Y \sim P(1.5, 1)$ on the left side, and excess wealth functions of $X \sim P(2.5, 2.5)$ (continuous line) and $Y \sim P(1.5, 1)$ (dashed line) on the right side.*

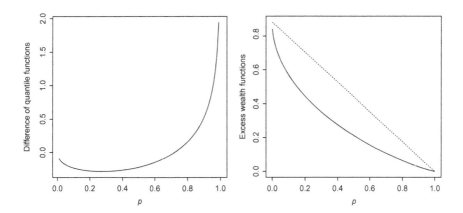

Figure 2.17 Difference of the quantile functions of $X \sim W(1,2)$ and $Y \sim W(\sqrt{\pi}/2,1)$ on the left side, and excess wealth functions of $X \sim W(1,2)$ (continuous line) and $Y \sim W(\sqrt{\pi}/2,1)$ (dashed line) on the right side.

$$E[X] = \alpha_1 \Gamma\left(\frac{\beta_1 + 1}{\beta_1}\right) \le \alpha_2 \Gamma\left(\frac{\beta_2 + 1}{\beta_2}\right) = E[Y],$$

then $X \le_{\text{ew}} Y$, but $X \not\le_{\text{disp}} Y$ or $X \not\ge_{\text{disp}} Y$. Figure 2.17 shows a particular case of this situation.

If the supports are intervals and the distribution functions are differentiable, it is easy to see that the previous corollary can be written as follows.

Corollary 2.6.25. *Let X and Y be two random variables with interval supports, with differentiable distribution functions F and G, density functions f and g, respectively, and finite means such that $\lim_{p \to 0^+}(G^{-1}(p) - F^{-1}(p)) \le E[Y] - E[X]$. Let us assume that there exists a value $p_0 \in (0,1)$ such that $g(G^{-1}(p)) \ge f(F^{-1}(p))$, for all $p \in (0,p_0)$ and $g(G^{-1}(p)) \le f(F^{-1}(p))$, for all $p \in [p_0, 1)$, then*

$$X \le_{\text{ew}} Y.$$

Let us apply this result to compare two Govindarajulu distributions.

Example 2.6.26. Let $X \sim G(\beta_1, \sigma_1, \theta_1)$ and $Y \sim G(\beta_2, \sigma_2, \theta_2)$ with quantile functions F^{-1} and G^{-1}, respectively. We study the crossing points of the functions $f(F^{-1}(p))/(1-p)$ and $g(G^{-1}(p))/(1-p)$, which is obviously equivalent to the sufficient condition of the previous corollary. It is not difficult to see that there is a crossing point at

$$p_0 = \left(\frac{(\beta_2 + 1)\beta_2 \sigma_2}{(\beta_1 + 1)\beta_1 \sigma_1} \right)^{\frac{1}{\beta_1 - \beta_2}} \in (0, 1),$$

in the sense stated in the Corollary 2.6.25, whenever the following conditions hold:

$$\beta_1 < \beta_2, \tag{2.43}$$

$$(\beta_2 + 1)\beta_2 \sigma_2 > (\beta_1 + 1)\beta_1 \sigma_1 \tag{2.44}$$

To apply Corollary 2.6.25, it is also required that both

$$\lim_{p \to 0^+} (G^{-1}(p) - F^{-1}(p)) = \theta_2 - \theta_1 \leq 0, \tag{2.45}$$

and

$$E[X] = \theta_1 + \sigma_1 \left(1 - \frac{\beta_1}{\beta_1 + 2} \right) \leq \theta_2 + \sigma_2 \left(1 - \frac{\beta_2}{\beta_2 + 2} \right) = E[Y]. \tag{2.46}$$

Notice that conditions (2.43), (2.45) and (2.46) imply (2.44). To sum up, if (2.43), (2.45) and (2.46) hold, then $X \leq_{\text{ew}} Y$, but $X \not\leq_{\text{disp}} Y$ or $X \not\geq_{\text{disp}} Y$.

Next, some preservation results for the excess wealth order are given. First, the preservation result under convergence is introduced.

Theorem 2.6.27. *Let $\{X_n\}_{n \in \mathbb{N}}$ and $\{Y_n\}_{n \in \mathbb{N}}$ be two sequences of random variables such that X_n converges in distribution to X and Y_n converges in distribution to Y, where X and Y are two continuous random variables with strictly increasing distribution functions. If $X_n \leq_{\text{ew}} Y_n$, for all $n \in \mathbb{N}$, and $E[(X_n)_+] \to E[(X)_+]$ and $E[(Y_n)_+] \to E[(Y)_+]$, then*

$$X \leq_{\text{ew}} Y.$$

Proof. Let us denote by F_n and G_n the distribution function of X_n and Y_n, respectively. From the assumptions and the arguments provided in Theorem 2.3.11, we see that $E[(X_n - t)_+] \to E[(X - t)_+]$. Now, according to the proof of Theorem 2.6.11, it holds that $F_n^{-1}(p) \to F^{-1}(p)$, for all $p \in (0, 1)$ and analogously for G_n. Therefore, we see that

$$\int_{F_n^{-1}(p)}^{+\infty} \overline{F}_n(x) \, dx \to \int_{F^{-1}(p)}^{+\infty} \overline{F}(x) \, dx$$

and analogously for Y, hence the result follows, since $X_n \leq_{\text{ew}} Y_n$, for all $n \in \mathbb{N}$. $\qquad \square$

The excess wealth order is also location free, that means, if $X \leq_{\text{ew}} Y$, then $X + c \leq_{\text{ew}} Y$ and $X \leq_{\text{ew}} Y + c$, for all $c \in \mathbb{R}$. Moreover, the excess wealth order is preserved under any increasing convex transformation. Before stating and proving this result, the following lemma provided by Barlow and Proschan [10] is required.

Lemma 2.6.28. *Let W be a measure on the interval (a, b), not necessarily non-negative, where $-\infty \leq a < b \leq +\infty$. Let h be a non-negative function defined on (a, b). If $\int_t^b dW(u) \geq 0$, for all $t \in (a, b)$, and h is increasing, then*

$$\int_t^b h(u)dW(u) \geq 0, \quad \text{for all } t \in (a, b).$$

Let us now introduce a preservation result under increasing convex transformations.

Theorem 2.6.29. *Let X and Y be two continuous random variables with finite left extreme points of their supports such that $l_X \leq l_Y$. If $X \leq_{\text{ew}} Y$, then*

$$\phi(X) \leq_{\text{ew}} \phi(Y), \text{ for all real valued increasing convex function } \phi.$$

Proof. Due to ϕ being increasing convex, then ϕ is strictly increasing. Let us assume that ϕ is also differentiable, without loss of generality. It is easy to see that the excess wealth transform of $\phi(X)$ is given by

$$W_{\phi(X)(p)} = \int_0^{+\infty} \overline{F}(z + F^{-1}(p))\phi'(z + F^{-1}(p))\, dz,$$

and analogously for $\phi(Y)$. Therefore, we have to prove that

$$\int_0^{+\infty} \overline{G}\left(z + G^{-1}(p)\right)\phi'\left(z + G^{-1}(p)\right) dz \tag{2.47}$$

$$\geq \int_0^{+\infty} \overline{F}\left(z + F^{-1}(p)\right)\phi'\left(z + F^{-1}(p)\right) dz,$$

for all $p \in (0, 1)$. We have to distinguish two cases.

(a) Assume $G^{-1}(p) \geq F^{-1}(p)$. By the monotonicity of ϕ', we see that

$$\int_0^{+\infty} \left[\overline{G}\left(z + G^{-1}(p)\right)\phi'\left(z + G^{-1}(p)\right) \right. \tag{2.48}$$

$$\left. -\overline{F}\left(z + F^{-1}(p)\right)\phi'\left(z + F^{-1}(p)\right)\right] dz$$

$$\geq \int_0^{+\infty} \left[\overline{G}\left(z + G^{-1}(p)\right) - \overline{F}\left(z + F^{-1}(p)\right)\right]\phi'\left(z + F^{-1}(p)\right) dz.$$

From (2.38), the hypothesis $X \leq_{\text{ew}} Y$ is equivalent to

$$\int_x^{+\infty} \left[\overline{G}\left(z + G^{-1}(p)\right) - \overline{F}\left(z + F^{-1}(p)\right) \right] \, dz \geq 0, \quad \text{for all } x \geq 0.$$

Since $\phi'\left(x + F^{-1}(p)\right)$ is monotone and positive, applying Lemma 2.6.28, we see that

$$\int_0^{+\infty} \left[\overline{G}\left(z + G^{-1}(p)\right) - \overline{F}\left(z + F^{-1}(p)\right) \right] \phi'\left(z + F^{-1}(p)\right) \, dz \geq 0,$$

Combining the previous inequality with (2.48), we get (2.47).

(b) Assume $G^{-1}(p) < F^{-1}(p)$.

Let us denote $p_1 = \sup\left\{u < p \mid G^{-1}(u) \geq F^{-1}(u)\right\}$, which exists by the hypothesis on the left extreme points of the supports. Then, we have

$$W_{\phi(Y)(p)} = \int_{G^{-1}(p_1)}^{+\infty} \overline{G}(z)\,\phi'(z)\,dz - \int_{G^{-1}(p_1)}^{G^{-1}(p)} \overline{G}(z)\,\phi'(z)\,dz$$

$$\geq \int_{F^{-1}(p_1)}^{+\infty} \overline{F}(z)\,\phi'(z)\,dz - \int_{G^{-1}(p_1)}^{G^{-1}(p)} \overline{G}(z)\,\phi'(z)\,dz,$$

where the inequality follows from (2.48). If we prove that

$$\int_{G^{-1}(p_1)}^{G^{-1}(p)} \overline{G}(z)\,\phi'(z)\,dz \leq \int_{F^{-1}(p_1)}^{F^{-1}(p)} \overline{F}(z)\,\phi'(z)\,dz,$$

we get the result. Notice that $\left[G^{-1}(p_1), G^{-1}(p) \right] \subseteq \left[F^{-1}(p_1), F^{-1}(p) \right]$, thus it is enough to prove that

$$\int_{G^{-1}(p_1)}^{G^{-1}(p)} \overline{G}(z)\,\phi'(z)\,dz \leq \int_{G^{-1}(p_1)}^{G^{-1}(p)} \overline{F}(z)\,\phi'(z)\,dz,$$

which follows if we prove that $\overline{G}(x) \leq \overline{F}(x)$, for all $x \in \left[G^{-1}(p_1), G^{-1}(p) \right]$. If $F(x) \leq p_1$, the inequality is obvious; otherwise, from the hypothesis $G^{-1}(p) < F^{-1}(p)$, we have $x < F^{-1}(p)$ and, consequently, $F(x) < p$. From the definition of p_1, we get $G^{-1}(F(x)) < F^{-1}(F(x))$. Since $F^{-1}(F(x)) < x$, we get $F(x) \leq G(x)$.

$\square$

To conclude with the preservation results, the theorem on the preservation of the excess wealth order under convolution is established. The proof is omitted here but can be seen in Ref. [74].

Theorem 2.6.30. *Let* $X_1, \ldots, X_n$ *and* $Y_1, \ldots, Y_n$ *be two sets of independent random variables such that* X_i *and* Y_i *are ILR and* $X_i \leq_{ew} Y_i$, *for all* $i = 1, \ldots, n$, *then*

$$\sum_{i=1}^{n} X_i \leq_{ew} \sum_{i=1}^{n} Y_i.$$

Finally, as far as graphical procedures is concerned, several possibilities exist to check the dispersive and excess wealth orders. First, we focus on the dispersive order, which can be verified through the $Q - Q$ plot. Since $X \leq_{disp} Y$ is equivalent to require $G^{-1}(p) - F^{-1}(p)$ to be increasing in p, the dispersive order holds if, and only if, the slope of every pair of points belonging to the $Q - Q$ plot is greater or equal than one [75]. Getting back on track with the example considered for the graphical procedure to check the stochastic order, let us consider now the data set corresponding to the thymic lymphoma death. The data set from the conventional laboratory environment and the germ free environment will be denoted by TCLE and TGFE, respectively. Figure 2.18 shows the empirical $Q-Q$ plot and suggests that the dispersive order does not hold.

Let us focus now on the excess wealth order. In this case, there are two possibilities to check this criterion graphically. On the one hand, it is possible to verify this order through the $Q - Q$ plot from the sufficient condition given by Theorem 2.6.20. Figure 2.18 suggests that a point p_0

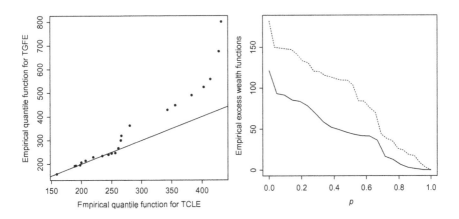

Figure 2.18 Empirical $Q - Q$ plot for TCLE and TGFE on the left side, and empirical excess wealth functions for TCLE (continuous line) and TGFE (dashed line) on the right side.

exists such that $G^{-1}(p) - F^{-1}(p) \leq E[Y] - E[X]$, for all $p \leq p_0$ (since the former points are on the diagonal $x = y$ and, consequently, the difference $G^{-1}(p) - F^{-1}(p)$ is closed to zero), and the slopes of the lines joining any pair of points of the $Q - Q$ plot are greater or equal than one, for all $p_0 < p < q$. Therefore, the plot suggests that the excess wealth order holds under the additional assumption of ordered means. Notice that it is also possible to check the sufficient condition stated by Corollary 2.6.22, that means, finding a point p_0 such that the slopes are smaller than one, for all $p < q < p_0$ and greater than one, for all $p_0 < p < q$. On the other hand, it is possible to plot the empirical excess wealth functions, as we can see in Figure 2.18. Once more, the plot suggests that the excess wealth order holds.

2.7 CONCENTRATION ORDERS

Closely related to the comparison of two random variables in terms of their variability, several criteria exist to compare random variables in terms of their concentration. This topic has great interest to compare income distributions in economy, where the basis of the comparison is the inequality among population. In this context, the more concentration of the incomes, the more inequality in the distribution of the total income among the population. Probably the most important criterion in this kind of comparison is the star-shaped order. Let us introduce the formal definition.

Definition 2.7.1. Given two random variables X and Y, we say that X is smaller than Y in *the star-shaped order*, denoted by $X \leq_\star Y$, if

$$F^{-1}(p)G^{-1}(q) \geq F^{-1}(q)G^{-1}(p), \quad \text{for all } 0 < p < q < 1.$$

If the random variables are non-negative, then $X \leq_\star Y$ is equivalent to

$$\frac{G^{-1}(p)}{F^{-1}(p)} \text{ be increasing in } p \in (0, 1). \tag{2.49}$$

In this section, we shall focus on non-negative random variables for discussion of the criteria, since the main application of theses orders is given in such case.

Furthermore, if the support of X is an open interval, then $X \leq_\star Y$ if, and only if, $G^{-1}(F(x))$ is star-shaped in x, that is, if

$$\frac{G^{-1}(F(x))}{x} \text{ is increasing in } x \geq 0. \tag{2.50}$$

The star-shaped order is preserved under positive scale changes, that is, if $X \leq_\star Y$, then $cX \leq_\star Y$ and $X \leq_\star cY$, for all $c > 0$. Unfortunately, not many properties have been developed for the star-shaped order. One of most useful known properties is the following one, which characterizes the star-shaped order in terms of the dispersive order.

Theorem 2.7.2. *Let X and Y be two non-negative random variables. Then, $X \leq_\star Y$ if, and only if, $\log(X) \leq_{\mathrm{disp}} \log(Y)$.*

Proof. Let F^{-1} $[G^{-1}]$ and F_l^{-1} $[G_l^{-1}]$ denote the quantile functions of X [Y] and $\log(X)$ [$\log(Y)$], respectively. From the property v) of the quantile function, we have $F_l^{-1}(p) = \log(F^{-1}(p))$ and $G_l^{-1}(p) = \log(G^{-1}(p))$, then

$$G_l^{-1}(p) - F_l^{-1}(p) = \log\left(\frac{G^{-1}(p)}{F^{-1}(p)}\right),$$

hence the result follows from (2.49).   □

Through this theorem, it is possible to provide several results for the star-shaped order from the previous ones for the dispersive order. In particular, the next result follows from the previous characterization and Theorem 2.6.4.

Theorem 2.7.3. *Let X and Y be two non-negative random variables with quantile functions F^{-1} and G^{-1}, respectively. Then, $X \leq_\star Y$ if, and only if,*

$$\left(\frac{X - F^{-1}(p)}{F^{-1}(p)}\right)_+ \leq_{\mathrm{st}} \left(\frac{Y - G^{-1}(p)}{G^{-1}(p)}\right)_+, \quad \textit{for all } p \in (0,1).$$

Through Theorem 2.7.2, a set of sufficient conditions is also provided for the star-shaped order in terms of the density functions.

Theorem 2.7.4. *Let X and Y be two non-negative continuous random variables with density functions f and g, respectively. If $S^-(g(\cdot) - cf(\cdot c)) \leq 2$ with the sign sequence $+, -, +$ in case of equality, for all $c \geq 0$, then*

$$X \leq_\star Y.$$

Proof. Let f_l and g_l denote the density functions of $\log(X)$ and $\log(Y)$, respectively. From Theorem 2.6.9, the condition $S^-(g_l(\cdot) - f_l(\cdot - c)) \leq 2$ with the sign sequence $+, -, +$ in case of equality, for all $c \in \mathbb{R}$ implies $\log(X) \leq_{\mathrm{disp}} \log(Y)$ or, equivalently, $X \leq_\star Y$, by Theorem 2.7.2. Since $S^-(g_l(\cdot) - f_l(\cdot - c))$, for all $c \in \mathbb{R}$ is equivalent to $S^-(g(\cdot) - cf(\cdot c))$, for all $c > 0$, we get the result.   □

This result has particular interest when no closed expression is available for the distribution function and/or the quantile function. Next, we give a particular scenario where the previous condition holds [76].

Theorem 2.7.5. *Let X and Y be two absolutely continuous random variables with common support $(0, +\infty)$ and density functions f and g, respectively. Let us consider the family of functions $h_c(x) = f(x) / (c\, g\,(cx))$, for all $c, x > 0$. If*

(i) $\lim_{x \to 0+} h_c(x) < 1$ or $\lim_{x \to +\infty} h_c(x) < 1$ and
(ii) $h_c(x)$ is monotone or has just one relative extreme,

then,

$$X \leq_\star Y.$$

Proof. The result follows trivially observing that the conditions of Theorem 2.7.4 are implied by hypotheses (i) and (ii).   □

Next, we give some applications of the previous theorem. First, two generalized gamma distributions are compared in the star-shaped order.

Example 2.7.6. Let $X \sim GG(a_1, b_1, p_1)$ and $Y \sim GG(a_2, b_2, p_2)$. Since the star-shaped order is scale invariant, we can assume that $b_1 = b_2 = 1$. According to the notation in Theorem 2.7.5, it is easy to see that $S^-(h_c(x)) = S^-(l_c(x))$, where

$$l_c(x) = (a_1 p_1 - a_2 p_2) \log(x) - x^{a_1} + (cx)^{a_2}.$$

Taking derivatives, we get

$$S^-(l_c'(x)) = S^-(-a_1 x^{a_1} + c^{a_2} a_2 x^{a_2} + (a_1 p_1 - a_2 p_2)). \qquad (2.51)$$

If we consider the second derivative, we see that

$$S^-(l_c''(x)) = S^-(-a_1^2 x^{a_1 - a_2} + c^{a_2} a_2^2) \leq 1, \quad \text{for all } x > 0,$$

hence $l_c'(x)$ is unimodal or monotone. Assuming $a_1 p_1 \geq a_2 p_2$ and $a_1 \geq a_2$, if we take limits as x tends to 0 and ∞ in the expression in (2.51), we get $a_1 p_1 - a_2 p_2 \geq 0$ and $-\infty$, respectively. Consequently, we see that $S^-(l_c'(x)) \leq 1$ with the sign sequence $+, -$ in case of equality. Therefore, $h_c(x)$ has at most one relative extreme. To sum up, if $a_1 \geq a_2$ and $a_1 p_1 \geq a_2 p_2$, then $X \leq_\star Y$.

This family contains as a particular case the Weibull distribution. Figure 2.19 shows particular cases of this situation.

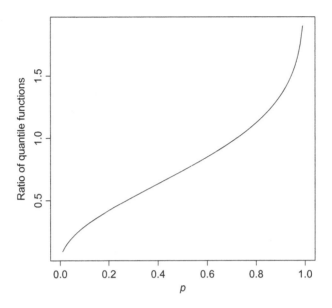

Figure 2.19 Ratio of the quantile functions of $X \sim W(1,2)$ and $Y \sim W(\sqrt{\pi}/2, 1)$.

Next, another set of sufficient conditions is provided for the star-shaped order when the random variables can be obtained as a transformation of a certain random variable.

Lemma 2.7.7. *Let X be a non-negative random variable (not degenerate at 0) with distribution function F, and h_1 and h_2 be two strictly increasing continuous functions, defined on $[0, +\infty)$, such that $h_1(x) > 0$ and $h_2(x) > 0$, for all $x > 0$. If $h_2(x)/h_1(x)$ is increasing in $x \in (0, +\infty)$, then*

$$h_1(X) \leq_\star h_2(X).$$

Proof. The proof follows from Theorems 2.6.13 and 2.7.2 observing that the quantile functions of $h_1(X)$ and $h_2(X)$ are given by $h_1(F^{-1}(p))$ and $h_2(F^{-1}(p))$, respectively, from the property v) of the quantile function. □

Next, the previous result combined with Theorem 2.7.5 is applied to compare two generalized type II beta distributions.

Example 2.7.8. Let $X \sim GB2(a_1, b_1, p_1, q_1)$ and $Y \sim GB2(a_2, b_2, p_2, q_2)$. Assume $b_1 = b_2 = 1$ as in the previous example. According to the notation of Theorem 2.7.5 and denoting by $A = \beta(p_2, q_2) / (c^{ap_2} \beta(p_1, q_1))$, we see that

$$h_c(x) = Ax^{a(p_1-p_2)} \frac{(1+(cx)^a)^{p_2+q_2}}{(1+x^a)^{p_1+q_1}}.$$

Taking derivatives, it is easy to see that

$$S^-(h_c'(x))$$

$$= S^-(-c^a(q_1-q_2)x^{2a} + [c^a(p_1+q_2) - (p_2+q_1)]x^a + (p_1-p_2)).$$

Replacing $x^a = t$ in the previous expression, we get a parabola and, therefore, $S^-(h_c'(x)) \le 2$. If we assume $a_1 p_1 \ge a_2 p_2$ and $a_1 q_1 \ge a_2 q_2$ (which is equivalent to $p_1 \ge p_2$ and $q_1 \ge q_2$ in our particular discussion), then the parabola has one positive root at most, hence $S^-(h_c'(x)) \le 1$, or equivalently, $h_c(x)$ has one relative extreme at most. Furthermore, $\lim_{x\to 0^+} h_c(x) = 0$. This discussion proves that the conditions of Theorem 2.7.5 are satisfied.

Let us suppose now $a_1 > a_2$ and let us consider the random variable $Z \sim GB2(1, p_1, q_1)$ and the functions $g, h : \mathbb{R}^+ \mapsto \mathbb{R}^+$ defined by $g(z) = z^{1/a_1}$ y $h(z) = z^{1/a_2}$. Lemma 2.7.7 ensures that $g(Z) \le_\star h(Z)$. It is not difficult to see that $Z^{\frac{1}{a}} \sim GB2(a, p_1, q_1)$ and, therefore, $g(Z) =_{st} X$ and $h(Z) \sim GB2(a_2, p_1, q_1)$. Finally, from the first part of the discussion, we see that $h(Z) \le_\star Y$, and the result follows from the transitivity of the star-shaped order. To sum up, if $a_1 \ge a_2$, $a_1 p_1 \ge a_2 p_2$ and $a_1 q_1 \ge a_2 q_2$, then $X \le_\star Y$.

Analogously to the dispersive orders, there are many ways of comparing random variables in terms of their concentration. For instance, it is possible to verify the expected proportional shortfall order in those situations where the star-shaped order does not hold. This criterion is defined through the comparison of the expected proportional shortfall functions. Let us see its formal definition. The upcoming results and some additional results can be found in Refs. [13, 73].

Definition 2.7.9. Given two non-negative random variables X and Y with finite means, we say that X is smaller than Y in the *expected proportional shortfall order*, denoted by $X \le_{ps} Y$, if

$$EPS_X(p) \le EPS_Y(p), \quad \text{for all } p \in (0, 1).$$

The definition of this criterion and also some of the upcoming results can be extended to non-necessarily non-negative random variables, considering the condition

$$\text{EPS}_X(p) \le \text{EPS}_Y(p), \quad \text{for all } p \in \{q \in (0,1)|F^{-1}(q), G^{-1}(q) \ne 0\},$$

where F^{-1} and G^{-1} denote the quantile functions of X and Y, respectively.

Clearly, the expected proportional shortfall order is scale invariant. From the different expressions of the expected proportional shortfall functions (see Section 1.2), we have $X \le_{\text{ps}} Y$ if, and only if, one of the following equivalent conditions holds:

(i) $\frac{W_X(p)}{\text{VaR}[X;p]} \le \frac{W_Y(p)}{\text{VaR}[Y;p]}$, for all $p \in (0,1)$,

(ii)
$$\frac{\text{TVaR}[X;p]}{\text{VaR}[X;p]} \le \frac{\text{TVaR}[Y;p]}{\text{VaR}[Y;p]}, \quad \text{for all } p \in (0,1). \tag{2.52}$$

In the continuous case, we have $X \le_{\text{ps}} Y$ if, and only if, one of the following equivalent conditions holds:

(iii) $\frac{\text{CTE}[X;p]}{\text{VaR}[X;p]} \le \frac{\text{CTE}[Y;p]}{\text{VaR}[Y;p]}$, for all $p \in (0,1)$,

(iv) $\frac{\text{CVaR}[X;p]}{\text{VaR}[X;p]} \le \frac{\text{CVaR}[Y;p]}{\text{VaR}[Y;p]}$, for all $p \in (0,1)$.

These characterizations also highlight that the expected proportional shortfall order compares, in a unified way, different measures of risks, but proportionally to the VaR. From the interpretations of the expected proportional shortfall function, the ps order also compares the relative deprivation among two populations.

Let us introduce two characterizations of the expected proportional shortfall order.

Theorem 2.7.10. *Let X and Y be two non-negative random variables with quantile functions F^{-1} and G^{-1}, respectively, and interval supports. Then, $X \le_{\text{ps}} Y$ if, and only if, one of the following equivalent conditions holds:*

(i)
$$\frac{\text{TVaR}[Y;p]}{\text{TVaR}[X;p]} \text{ is increasing in } p \in (0,1). \tag{2.53}$$

(ii)
$$\frac{\int_q^1 F^{-1}(u)\, du}{F^{-1}(p)} \le \frac{\int_q^1 G^{-1}(u)\, du}{G^{-1}(p)}, \quad \text{for all } 0 < p \le q < 1. \tag{2.54}$$

Proof. It is clear that (i) and (ii) are equivalents. Furthermore, taking derivatives, (2.53) is equivalent to condition (2.52). $\square$

From the previous characterizations, it is easy to see that the expected proportional shortfall order is weaker than the star-shaped order, that is,

$$X \leq_\star Y \Rightarrow X \leq_{ps} Y.$$

The combination of Theorem 2.3.3 with (2.54) provides the following characterization of the expected proportional shortfall order.

Theorem 2.7.11. *Let X and Y be two non-negative random variables with quantile functions F^{-1} and G^{-1}, respectively, interval supports and finite means. Then $X \leq_{ps} Y$ if, and only if,*

$$\left(\frac{X - F^{-1}(p)}{F^{-1}(p)}\right)_+ \leq_{icx} \left(\frac{Y - G^{-1}(p)}{G^{-1}(p)}\right)_+, \quad \text{for all } p \in (0, 1).$$

Analogously to the excess wealth order, the verification of the expected proportional shortfall order requires the evaluation of certain incomplete integrals of the quantile functions, which do not always have closed expressions either. Next, we give a set of sufficient conditions in terms of the quantile functions to overcome this difficulty in many situations as possible [73].

Theorem 2.7.12. *Let X and Y be two random variables with quantile functions F^{-1} and G^{-1}, respectively, and finite means such that $\lim_{p \to 0^+} G^{-1}(p)/F^{-1}(p) \leq E[Y]/E[X]$. If a point $p_0 \in (0, 1)$ exists such that $G^{-1}(p)/F^{-1}(p)$ is decreasing in $p \in (0, p_0)$, and $G^{-1}(p)/F^{-1}(p)$ is increasing in $p \in [p_0, 1)$, then*

$$X \leq_{ps} Y.$$

Proof. Let us consider $p \in [p_0, 1)$ and the random variables $\left(\frac{X - F^{-1}(p)}{F^{-1}(p)}\right)_+$ and $\left(\frac{Y - G^{-1}(p)}{G^{-1}(p)}\right)_+$. According to (1.7), the quantile function associated to $\left(\frac{X - F^{-1}(p)}{F^{-1}(p)}\right)_+$ is given by

$$F_*^{-1}(q; p) = \left(\frac{F^{-1}(q) - F^{-1}(p)}{F^{-1}(p)}\right)_+, \tag{2.55}$$

and analogously to $\left(\frac{Y - G^{-1}(p)}{G^{-1}(p)}\right)_+$. Since $G^{-1}(p)/F^{-1}(p)$ is increasing in $p \in [p_0, 1)$, it holds trivially from (2.55) that

$$\left(\frac{X - F^{-1}(p)}{F^{-1}(p)}\right)_{+} \leq_{st} \left(\frac{Y - G^{-1}(p)}{G^{-1}(p)}\right)_{+}, \quad \text{for all } p \in [p_0, 1),$$

and, therefore,

$$\text{EPS}_X(p) \leq \text{EPS}_Y(p), \quad \text{for all } p \in [p_0, 1), \tag{2.56}$$

from the preservation of the means by the stochastic order. Let us consider now $p \in (0, p_0)$ and the random variable $\left(\frac{X - F^{-1}(p)}{F^{-1}(p)}\right)_{-}$ with quantile function given by

$$F_{-\star}^{-1}(q; p) = \left(\frac{F^{-1}(q) - F^{-1}(p)}{F^{-1}(p)}\right)_{-}, \quad \text{for all } q \in (0, 1),$$

and analogously to $\left(\frac{Y - G^{-1}(p)}{G^{-1}(p)}\right)_{-}$. Since $G^{-1}(p)/F^{-1}(p)$ is decreasing in $p \in (0, p_0)$, then

$$\left(\frac{X - F^{-1}(p)}{F^{-1}(p)}\right)_{-} \geq_{st} \left(\frac{Y - G^{-1}(p)}{G^{-1}(p)}\right)_{-}, \quad \text{for all } p \in (0, p_0),$$

and, by (2.2), this implies that

$$E\left[\left(\frac{X - F^{-1}(p)}{F^{-1}(p)}\right)_{-}\right] \geq E\left[\left(\frac{Y - G^{-1}(p)}{G^{-1}(p)}\right)_{-}\right], \quad \text{for all } p \in (0, p_0). \tag{2.57}$$

Furthermore, from the hypothesis, we see that

$$\frac{G^{-1}(p)}{F^{-1}(p)} \leq \lim_{p \to 0+} \frac{G^{-1}(p)}{F^{-1}(p)} \leq \frac{E(Y)}{E(X)}, \quad \text{for all } p \in (0, p_0). \tag{2.58}$$

Combining the equality $(x)_{+} = x - (x)_{-}$ with (2.58) and (2.57), it is verified that

$$E\left[\left(\frac{X - F^{-1}(p)}{F^{-1}(p)}\right)_{+}\right] \leq E\left[\left(\frac{Y - G^{-1}(p)}{G^{-1}(p)}\right)_{+}\right], \quad \text{for all } p \in (0, p_0). \tag{2.59}$$

The result follows combining (2.56) and (2.59).   □

Let us apply this theorem to compare two Govindarajulu distributions.

Example 2.7.13. Let $X \sim G(\beta_1, \sigma_1, \theta_1)$ and $Y \sim G(\beta_2, \sigma_2, \theta_2)$ with quantile functions F^{-1} and G^{-1}, respectively. We assume $\sigma_1 = \sigma_2 = 1$ because of the preservation under scale changes. Denoting by F^{-1} and G^{-1}

the quantile functions of X and Y, we will study the monotonicity of the ratio $G^{-1}(p)/F^{-1}(p)$ through $S^-((\log(G^{-1}(p)) - \log(F^{-1}(p)))') = S^-(h(p))$, where

$$h(p) = \theta_1\beta_2(\beta_2 + 1) - \theta_2\beta_1(\beta_1 + 1)p^{\beta_1-\beta_2} +$$
$$p^{\beta_1}(\beta_1 - \beta_2)[\beta_1\beta_2 p - (\beta_1 + 1)(\beta_2 + 1)],$$

for all $p \in (0, 1)$. Since

$$h'(p) = p^{\beta_1-1}[\beta_1(\beta_1 + 1)(\beta_2 - \beta_1)\left[\theta_2 p^{-\beta_2} + \beta_2(1 - p) + 1\right] > 0,$$

for all $p \in (0, 1)$, whenever $\beta_1 < \beta_2$, then $h(p)$ is increasing in $p \in (0, 1)$ and, consequently, $S^-(h(p)) \leq 1$. Moreover, $\lim_{p\to 0+} h(p) = -\infty$, $\lim_{p\to 1-} h(p) = (\theta_1 + 1)\beta_2(\beta_2 + 1) - (\theta_2 + 1)\beta_1(\beta_1 + 1)$ and

$$\lim_{p\to 0+} \frac{G^{-1}(p)}{F^{-1}(p)} = \begin{cases} \frac{\theta_2}{\theta_1} & \theta_1 > 0, \\ 0 & \theta_1 = \theta_2 = 0, \\ +\infty & \theta_1 = 0, \theta_2 > 0. \end{cases}$$

Consequently, we have to distinguish the following cases.

(a) $\theta_1 > 0$. The hypothesis $\lim_{p\to 0+} G^{-1}(p)/F^{-1}(p) \leq E[Y]/E[X]$ required in Theorem 2.7.12 is equivalent to the inequality $\theta_2(\beta_2+2) \leq \theta_1(\beta_1+2)$. Thus, if $\theta_2(\beta_2 + 2) \leq \theta_1(\beta_1 + 2)$ and $\beta_2 > \beta_1$, then $S^-(h(p)) = 1$ with the sign sequence $-,+$. Hence, $G^{-1}(p)/F^{-1}(p)$ has a minimum. To sum up, applying Theorem 2.7.12, if $\beta_1 \leq \beta_2$ and $\theta_2(\beta_2 + 2) \leq \theta_1(\beta_1 + 2)$, then $X \leq_{ps} Y$, but $X \not\leq_\star Y$ and $X \not\geq_\star Y$.

(b) $\theta_1 = \theta_2 = 0$. In this case, the hypothesis $\lim_{p\to 0+} G^{-1}(p)/F^{-1}(p) \leq E[Y]/E[X]$ is trivially verified. Again, if $\beta_2 > \beta_1$, then $S^-(h(p)) = 1$ with the sign sequence $-,+$ and, therefore, $G^{-1}(p)/F^{-1}(p)$ has a minimum. To sum up, applying Theorem 2.7.12, if $\beta_1 \leq \beta_2$, then $X \leq_{ps} Y$, but $X \not\leq_\star Y$ and $X \not\geq_\star Y$.

Notice that under the assumptions $\theta_1 = 0, \theta_2 > 0$, the condition $\lim_{p\to 0+} G^{-1}(p)/F^{-1}(p) \leq E[Y]/E[X]$ is no longer verified and, consequently, Theorem 2.7.12 cannot be applied.

Finally, a result on the preservation of the expected proportional shortfall order under convergence is given. The proof is similar to the one of Theorem 2.6.27 and it is omitted.

Theorem 2.7.14. *Let $\{X_n\}_{n\in\mathbb{N}}$ and $\{Y_n\}_{n\in\mathbb{N}}$ be two sequences of random variables such that X_n converges in distribution to a random variable X and*

Y_n *to a random variable* Y, *where* X *and* Y *are continuous random variables with interval supports. If* $X_n \leq_{ps} Y_n$, *for all* $n \in \mathbb{N}$ *and* $E\left[(X_n)_+\right] \to E[X_+]$ *and* $E\left[(Y_n)_+\right] \to E[Y_+]$, *then*

$$X \leq_{ps} Y.$$

As mentioned in Section 1.2, the main tool to assess concentration and inequality in income distributions is the Lorenz curve. The comparison of Lorenz curves leads to the following criterion to compare random variables.

Definition 2.7.15. Given two non-negative random variables X and Y, with finite means, we say that X is smaller than Y in the *Lorenz order*, denoted by $X \leq_L Y$ if

$$L_X(p) \geq L_Y(p), \quad \text{for all } p \in (0, 1).$$

The Lorenz order is scale invariant. The following characterization, which comes from Theorem 2.3.3, can be used to derive several properties of the Lorenz order which can be found in Refs. [2, 77].

Theorem 2.7.16. *Let* X *and* Y *be two non-negative random variables. Then* $X \leq_L Y$ *if, and only if,*

$$\frac{X}{E[X]} \leq_{cx} \frac{Y}{E[Y]},$$

or, equivalently, if

$$E\left[\phi\left(\frac{X}{E[X]}\right)\right] \leq E\left[\phi\left(\frac{Y}{E[Y]}\right)\right],$$

for all real valued convex function ϕ *such that the previous expectations exist.*

As a consequence of the previous result, the Lorenz order implies the ordering of the variation coefficients. Clearly, from the definition, the Lorenz order implies the ordering of the Gini indexes as well. Finally, we show that the Lorenz order is weaker than the expected proportional shortfall order.

Theorem 2.7.17. *Let* X *and* Y *be two non-negative random variables with interval supports. If* $X \leq_{ps} Y$, *then*

$$X \leq_L Y.$$

Proof. First, we observe that

$$\text{TVaR}[X; p] = \frac{E[X]}{1 - p} (1 - L_X(p)),$$

and analogously for $\text{TVaR}[Y; p]$. Therefore, from the characterization provided in (2.53), we see that $X \leq_{\text{ps}} Y$ if, and only if,

$$\frac{1 - L_Y(p)}{1 - L_X(p)} \text{ is increasing in } p \in (0, 1).$$

Since $L_X(0) = L_Y(0) = 0$, we see that

$$L_X(p) \geq L_Y(p), \quad \text{for all } p \in (0, 1).$$

$\square$

Repeatedly, we conclude this section introducing the available graphical procedures to check the criteria studied in this section. First, we focus on the star-shaped order, which can be checked through the $Q - Q$ plot. Since $X \leq_\star Y$ is equivalent to require $G^{-1}(p)/F^{-1}(p)$ to be increasing in p, it is obvious that the star-shaped order holds if, and only if, the $Q - Q$ plot is star-shaped with respect to $(0, 0)$. Consider the following data sets. The Secura Belgium Re (SB) data containing automobile claims for the period 1981–2001 gathered from several European insurance companies, which are at least as large as 1,200,000 Euro, and the SOA Group Medical Insurance Large Claims Database (SOA) containing all the claims exceeding 25,000 USD for the period 1991–1992. Both data sets can be found in Ref. [78]. Figure 2.20 suggests that the star-shaped order does not hold.

Next, we focus on the expected proportional shortfall order. In this case, there are two possibilities to check this criterion graphically. On the one hand, it is possible to verify this order through the $Q - Q$ plot from the sufficient conditions given by Theorem 2.7.12. If a point $p_0 \in (0, 1)$ exists such that the $Q - Q$ plot is anti star-shaped with respect to $(0, 0)$ from $(0, 0)$ until $(F^{-1}(p_0), G^{-1}(p_0))$, and star-shaped with respect to $(0, 0)$ from $(F^{-1}(p_0), G^{-1}(p_0))$ until $(1, 1)$. The plot on the left side of Figure 2.20 suggests that this sufficient condition is not verified and, therefore, we cannot conclude by this way. On the other hand, it is possible to plot the empirical expected proportional shortfall functions. The plot on the right side of Figure 2.20 suggests that the empirical expected proportional shortfall functions are ordered.

Finally, it is possible to check the Lorenz order for two data sets graphically through plotting the empirical Lorenz curves.

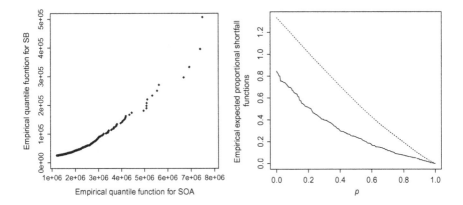

Figure 2.20 Empirical Q – Q plot for SOA and SB on the left side, and empirical expected proportional shortfall functions for SOA (continuous line) and SB (dashed line) on the right side. This figure was amended from Belzunce et al. [13].

2.8 THE TOTAL TIME ON TEST TRANSFORM ORDER

Another criterion to compare random variables is defined by means of the generalized ttt transforms (see Section 1.2). Let us see the formal definition of this order.

Definition 2.8.1. Given two random variables X and Y with finite means, we say that X is smaller than Y in the *total time on test transform order*, denoted by $X \leq_{ttt} Y$, if

$$T_X(p) \leq T_Y(p), \quad \text{for all } p \in (0, 1).$$

Recalling that $T_X(p) = E[\min\{X, F^{-1}(p)\}]$, and analogously for $T_Y(p)$, and from the interpretation of the random variable $\min\{X, F^{-1}(p)\}$ in both reliability and risk theory (see Section 1.2), clear interpretations of the ttt order arise in these contexts. Besides, this criterion can be used in every context where the main interest is to compare random variables in terms of lower tails. For instance, it can be applied in poverty studies related to income distributions where the main interest are lower incomes.

It is clear from Theorem 2.2.4 and (2.2) that

$$X \leq_{st} Y \Rightarrow X \leq_{ttt} Y.$$

Next, we provide a characterization of the ttt order which will be used later [15].

Theorem 2.8.2. *Let X and Y be two random variables with quantile functions F^{-1} and G^{-1}, respectively, interval supports and finite means such that $\lim_{p\to 0^+} \left(G^{-1}(p) - F^{-1}(p)\right)$ is finite. Then, $X \leq_{\text{ttt}} Y$ if, and only if,*

$$\frac{1}{1-p} \int_0^p (G^{-1}(u) - F^{-1}(u))\, du \text{ is increasing in } p \in (0, 1).$$

Proof. Since

$$T_X(p) = (1-p)F^{-1}(p) + \int_0^p F^{-1}(u)\, du, \quad \text{for all } p \in (0, 1),$$

the result follows by differentiation. $\qquad\qquad\qquad\qquad\qquad\qquad \square$

The upcoming theorem characterizes the ttt order in terms of the icv order, see Ref. [79].

Theorem 2.8.3. *Let X and Y be two random variables with continuous quantile functions F^{-1} and G^{-1}, respectively, interval supports and finite means such that $\lim_{p\to 0^+} \left(G^{-1}(p) - F^{-1}(p)\right)$ is finite. Then, $X \leq_{\text{ttt}} Y$ if, and only if,*

$$\min\{X, F^{-1}(p)\} \leq_{\text{icv}} \min\{Y, G^{-1}(p)\}, \quad \text{for all } p \in (0, 1).$$

Proof. From the preservation of the means by the icv order, we see that the condition

$$\min\{X, F^{-1}(p)\} \leq_{\text{icv}} \min\{Y, G^{-1}(p)\}, \quad \text{for all } p \in (0, 1), \qquad (2.60)$$

implies $X \leq_{\text{ttt}} Y$.

Let us prove the inverse implication. According to the notation in Theorem 2.2.4, let us denote by F_p^{-1} and G_p^{-1} the quantile functions of $\min\{X, F^{-1}(p)\}$ and $\min\{Y, G^{-1}(p)\}$, respectively. From Theorem 2.3.3 and (2.9), condition (2.60) is equivalent to

$$\int_0^q F_p^{-1}(u)\, du \leq \int_0^q G_p^{-1}(u)\, du, \quad \text{for all } p, q \in (0, 1).$$

From Theorem 2.8.2 and the fact that $\lim_{p\to 0^+} \left(G^{-1}(p) - F^{-1}(p)\right)$ is finite, we see that

$$\frac{1}{1-q} \int_0^q \left(G^{-1}(u) - F^{-1}(u)\right) du$$

$$\geq \lim_{p\to 0^+} \frac{1}{1-p} \int_0^p \left(G^{-1}(u) - F^{-1}(u)\right) du = 0.$$

Let us consider $q < p$. From (2.1) and the previous inequality, we have

$$\int_0^q G_p^{-1}(u)\,du - \int_0^q F_p^{-1}(u)\,du = \int_0^q G^{-1}(u)\,du - \int_0^q F^{-1}(u)\,du \geq 0.$$

Let us assume now $p \leq q < 1$. Again from (2.1), we see that

$$\begin{aligned}
H(p,q) &= \int_0^q G_p^{-1}(u)\,du - \int_0^q F_p^{-1}(u)\,du \\
&= \int_0^p \left(G^{-1}(u)\,du - F^{-1}(u) \right) du + (q-p)\left(G^{-1}(p) - F^{-1}(p) \right),
\end{aligned}$$

which is a straight line in $q \in [p, 1)$, where

$$H(p,p) = \int_0^p \left(G^{-1}(u)\,du - F^{-1}(u) \right) du \geq 0,$$

as we have seen previously. Furthermore

$$\lim_{q \to 1^-} H(q,p) = E[\min\{Y, G^{-1}(p)\}] - E[\min\{X, F^{-1}(p)\}] \geq 0,$$

since $X \leq_{\text{ttt}} Y$. Therefore, $H(p,q) \geq 0$, for all $q \in [p,1)$ and we get the result. $\qquad\square$

If the left extreme points of the support are finite such that $l_X \leq l_Y$, we have

$$X \leq_{\text{ttt}} Y \Rightarrow X \leq_{\text{icv}} Y,$$

from the proof of the previous characterization.

Repeatedly, this order requires the evaluation of incomplete integrals of the quantile functions, which is not possible in many situations. In such cases, one possibility is to check the stochastic order, which is a sufficient condition for the total on time transform order. However, there are multiple situations where the stochastic order does not hold. To overcome this difficulty, some sufficient conditions are provided for the total time on test transform order when the stochastic order does not hold [79].

Theorem 2.8.4. *Let X and Y be two random variables with quantile functions F^{-1} and G^{-1}, respectively and finite means such that $E[X] \leq E[Y]$. If a single crossing point $p_0 \in (0,1)$ exists among the quantile functions such that $G^{-1}(p) \geq F^{-1}(p)$, for all $p \in (0,p_0)$ and $G^{-1}(p) - F^{-1}(p)$ is decreasing in $p \in (p_0, 1)$, then*

$$X \leq_{\text{ttt}} Y.$$

Proof. Assume first $p \in (0, p_0)$. Combining (2.1) with the hypothesis $G^{-1}(p) \geq F^{-1}(p)$, for all $p \in (0, p_0)$, and (2.2), it holds that

$$E[\min\{X, F^{-1}(p)\}] \leq E[\min\{Y, G^{-1}(p)\}], \quad \text{for all } p \in (0, p_0). \quad (2.61)$$

Assume now $p \in [p_0, 1)$ and let us consider the random variables $(X - F^{-1}(p))_+$ and $(Y - G^{-1}(p))_+$. Since $G^{-1}(p) - F^{-1}(p)$ is decreasing in $p \in (p_0, 1)$, from (2.33), we get $(X - F^{-1}(p))_+ \geq_{st} (Y - G^{-1}(p))_+$, for all $p \in (p_0, 1)$ and, consequently,

$$E\left[(X - F^{-1}(p))_+\right] \geq E\left[(Y - G^{-1}(p))_+\right], \quad \text{for all } p \in (p_0, 1).$$

Combining the previous inequality with the hypothesis $E[X] \leq E[Y]$ and the fact that $\min\{x, t\} = x - (x - t)_+$, we have

$$E\left[\min\{X, F^{-1}(p)\}\right] \leq E\left[\min\{Y, G^{-1}(p)\}\right], \quad \text{for all } p \in (p_0, 1). \quad (2.62)$$

The result follows from (2.61) and (2.62). $\qquad\square$

Let us see next an application of this theorem to compare two normal and two Davies distributions.

Example 2.8.5. Let $X \sim N(\mu_1, \sigma_1^2)$ and $Y \sim N(\mu_2, \sigma_2^2)$ with quantile functions F^{-1} and G^{-1}. From Example 2.6.15, if $\sigma_1 > \sigma_2$, then $G^{-1}(p) - F^{-1}(p)$ is decreasing in $p \in (0, 1)$. From Taylor [83], and the fact that the stochastic order does not hold under the assumption $\sigma_1 > \sigma_2$, then a single crossing point $p_0 \in (0, 1)$ exists such that $G^{-1}(p) \geq F^{-1}(p)$, for all $p \in (0, p_0)$. Applying the previous theorem, if $\mu_1 \leq \mu_2$ and $\sigma_1 > \sigma_2$, then $X \leq_{ttt} Y$, but $X \not\leq_{st} Y$ or $X \not\geq_{st} Y$.

Example 2.8.6. Let $X \sim D(\lambda_1, \theta_1, C_1)$ and $Y \sim D(\lambda_2, \theta_2, C_2)$ with quantile functions F^{-1} and G^{-1}, respectively. In order to have finite means we assume $\theta_1, \theta_2 < 1$. From Example 2.6.21, if $\lambda_1 \geq \lambda_2$ and $\theta_1 \geq \theta_2$, a point $p_0 \in (0, 1)$ exists such that $G^{-1}(p) \geq F^{-1}(p)$, for all $p \in (0, p_0)$ and $G^{-1}(p) - F^{-1}(p)$ is decreasing in $p \in [p_0, 1)$. Thus, if $\lambda_1 \geq \lambda_2$, $\theta_1 \geq \theta_2$ and

$$E[X] = C_1 B(1 + \lambda_1, 1 - \theta_1) \leq E[Y] = C_2 B(1 + \lambda_2, 1 - \theta_2),$$

applying Theorem 2.8.4, we see that $X \leq_{\text{ttt}} Y$, but $X \nleq_{\text{st}} Y$ or $X \ngeq_{\text{st}} Y$. Since the Pareto family arises as a limit case of this model, given $X \sim P(a_1, k_1)$ and $Y \sim P(a_2, k_2)$ such that $a_1 < a_2$ and

$$E[X] = \frac{k_1 a_1}{a_1 - 1} \leq E[Y] = \frac{k_2 a_2}{a_2 - 1},$$

then $X \leq_{\text{ttt}} Y$, but $X \nleq_{\text{st}} Y$ or $X \ngeq_{\text{st}} Y$. Figure 2.21 shows a particular case of this situation.

In many situations, the difference of the quantiles is initially increasing and later decreasing. Additionally, if $\lim_{p \to 0^+} \left(G^{-1}(p) - F^{-1}(p) \right) \geq 0$ and $\lim_{p \to 1^-} \left(G^{-1}(p) - F^{-1}(p) \right) < 0$, the hypothesis on the difference of the quantiles required in Theorem 2.8.4 are satisfied. Hence, if $E[X] \leq E[Y]$, we see that $X \leq_{\text{ttt}} Y$. The Weibull family presents this behavior, as we will see next.

Example 2.8.7. Let $X \sim W(\alpha_1, \beta_1)$ and $Y \sim W(\alpha_2, \beta_2)$ with quantile functions F^{-1} and G^{-1}, respectively. From Example 2.6.24, $G^{-1}(p) - F^{-1}(p)$ has a maximum, whenever $\beta_1 < \beta_2$. Since both random variables are non-negative, then $\lim_{p \to 0^+}(G^{-1}(p) - F^{-1}(p)) = 0$ and it is easy to see that $\lim_{p \to 1^-}(G^{-1}(p) - F^{-1}(p)) = -\infty$. Therefore, if we assume $\beta_1 < \beta_2$ and

$$E[X] = \alpha_1 \Gamma \left(\frac{\beta_1 + 1}{\beta_1} \right) \leq \alpha_2 \Gamma \left(\frac{\beta_2 + 1}{\beta_2} \right) = E[Y],$$

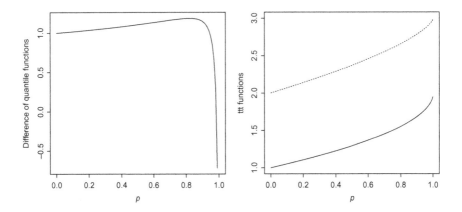

Figure 2.21 Difference of the quantile functions of $X \sim P(2,1)$ and $Y \sim P(3,2)$ on the left side, and ttt functions of $X \sim P(2,1)$ (continuous line) and $Y \sim P(3,2)$ (dashed line) on the right side.

by the previous comment and Theorem 2.8.4, we conclude $X \leq_{ttt} Y$, but $X \nleq_{st} Y$ or $X \ngeq_{st} Y$. Figure 2.22 shows a particular case of this situation.

To finish, we provide some preservation results for the ttt order. The proofs are similar to the ones for the excess wealth order and they are omitted. The first one is the preservation of the ttt order under increasing convex transformations [80].

Theorem 2.8.8. *Let X and Y be two continuous random variables with finite left extreme points such that $l_X = l_Y$. If $X \leq_{ttt} Y$, then*

$$\phi(X) \leq_{ew} \phi(Y), \text{ for all real valued increasing concave function } \phi.$$

The next result is on the preservation under convolutions [15].

Theorem 2.8.9. *Let $X_1, \ldots, X_n$ and $Y_1, \ldots, Y_n$ be two sets of independent random variables such that X_i and Y_i are ILR and $X_i \leq_{ttt} Y_i$, for all $i = 1, \ldots, n$. Then,*

$$\sum_{i=1}^{n} X_i \leq_{ttt} \sum_{i=1}^{n} Y_i.$$

Repeatedly, we conclude this section introducing the graphical procedures to check the total time on test transform order. Let us see two possibilities. On the one hand, it is possible to verify the order through the $Q - Q$ plot by means of Theorem 2.8.4. Figure 2.23 shows the empirical

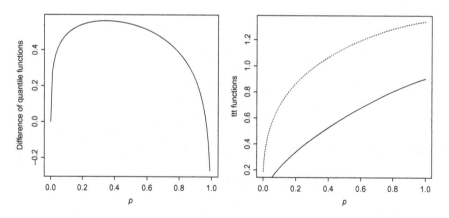

Figure 2.22 Difference of the quantile functions of $X \sim W(1, 1.5)$ and $Y \sim W(1.5, 3)$ on the left side, and ttt functions of $X \sim W(1, 1.5)$ (continuous line) and $Y \sim W(1.5, 3)$ (dashed line) on the right side.

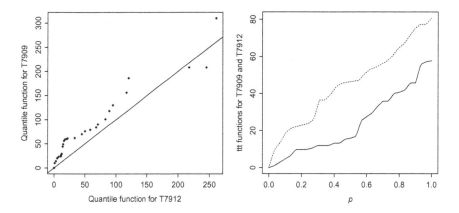

Figure 2.23 Empirical Q−Q plot for T7912 and T7909 on the left side, and empirical ttt functions for T7912 (continuous line) and T7909 (dashed line) on the right side.

$Q - Q$ plot for two data sets taken from Ref. [81]. This data represents the time between failure of the air conditioning of two subgroups of airplanes Boeing 720, the subgroup 7909 (T7909) and the subgroup 7912 (T7912). The conditions of Theorem 2.8.4 are not satisfied for these data sets and it is also worth mentioning that the stochastic order does not hold. The problem occurs after the cross among the quantile functions, since the slope of the lines joining the points are not always smaller than one. On the other hand, this criterion can be checked by plotting the ttt functions. Figure 2.23 suggests that the ttt order holds, since the empirical ttt functions seem to be ordered.

2.9 APPLICATIONS

In this section, a summary of the different comparisons of parametric models is given, as well as applications on the comparison of coherent systems, distorted distributions, and individual and collective risks models.

2.9.1 Comparison of parametric families of distributions
In the previous sections, a great many examples of two parametric distributions (belonging to the same model) have been ordered in the different studied criteria. In this section, a summary of such results, and some other results considered in the literature, is given. In particular, Tables 2.1 and 2.2 collect the results provided for the normal, lognormal, gamma, Weibull and Pareto family distributions, and some additional results provided by Lisek

Table 2.1 Comparison of some parametric continuous distributions in the likelihood ratio, the hazard rate, the stochastic, the dispersive, and the star-shaped orders

Family	Criterion				
	$X \leq_{lr} Y$	$X \leq_{hr} Y$	$X \leq_{st} Y$	$X \leq_{disp} Y$	$X \leq_{*} Y$
Normal $X \sim N(\mu_1, \sigma_1^2)$ $Y \sim N(\mu_2, \sigma_2^2)$	$\sigma_1 = \sigma_2$ $\mu_1 \leq \mu_2$	$\sigma_1 = \sigma_2$ $\mu_1 \leq \mu_2$	$\sigma_1 = \sigma_2$ $\mu_1 \leq \mu_2$	$\sigma_1 \leq \sigma_2$	
Lognormal $X \sim LN(\mu_1, \sigma_1^2)$ $Y \sim LN(\mu_2, \sigma_2^2)$	$\sigma_1 = \sigma_2$ $\mu_1 \leq \mu_2$	$\sigma_1 = \sigma_2$ $\mu_1 \leq \mu_2$	$\sigma_1 = \sigma_2$ $\mu_1 \leq \mu_2$	$\sigma_1 \leq \sigma_2$	$\sigma_1 \leq \sigma_2$
Gamma $X \sim G(\alpha_1, \beta_1)$ $Y \sim G(\alpha_2, \beta_2)$	$\alpha_1 \leq \alpha_2$ $\beta_1 \leq \beta_2$	$\alpha_1 \leq \alpha_2$ $\beta_1 \leq \beta_2$	$\alpha_1 \leq \alpha_2$ $\beta_1 \leq \beta_2$	$\alpha_1 \leq \alpha_2$ $\beta_1 \leq \beta_2$	$\alpha_1 \geq \alpha_2$
Weibull $X \sim W(\alpha_1, \beta_1)$ $Y \sim W(\alpha_2, \beta_2)$	$\alpha_1 \leq \alpha_2$ $\beta_1 = \beta_2$	$\alpha_1 \leq \alpha_2$ $\beta_1 = \beta_2$	$\alpha_1 \leq \alpha_2$ $\beta_1 = \beta_2$	$\alpha_1 \leq \alpha_2$ $\beta_1 = \beta_2$	$\beta_1 \geq \beta_2$
Pareto $X \sim P(a_1, k_1)$ $Y \sim P(a_2, k_2)$	$a_1 \geq a_2$ $k_1 \leq k_2$	$a_1 \geq a_2$ $k_1 \leq k_2$	$a_1 \geq a_2$ $k_1 \leq k_2$	$a_1 \geq a_2$ $a_1 k_2 \geq a_2 k_1$	$a_1 \geq a_2$

Table 2.2 Comparison of some parametric continuous distributions in the ttt, the mean residual life, the increasing convex, the excess wealth, and the expected proportional shortfall orders

Family	Criterion				
	$X \leq_{ttt} Y$	$X \leq_{mrl} Y$	$X \leq_{icx} Y$	$X \leq_{ew} Y$	$X \leq_{ps} Y$
Normal $X \sim N(\mu_1, \sigma_1^2)$ $Y \sim N(\mu_2, \sigma_2^2)$	$\sigma_1 > \sigma_2$ $\mu_1 \leq \mu_2$	$\sigma_1 < \sigma_2$ $\mu_1 \leq \mu_2$	$\sigma_1 < \sigma_2$ $\mu_1 \leq \mu_2$	$\sigma_1 \leq \sigma_2$	
Lognormal $X \sim LN(\mu_1, \sigma_1^2)$ $Y \sim LN(\mu_2, \sigma_2^2)$	$\sigma_1 = \sigma_2$ $\mu_1 \leq \mu_2$	$\sigma_1 < \sigma_2$ $\mu_1 \leq \mu_2$	$\sigma_1 < \sigma_2$ $\mu_1 \leq \mu_2$	$\sigma_1 \leq \sigma_2$	$\sigma_1 \leq \sigma_2$
Gamma $X \sim G(\mu_1, \beta_1)$ $Y \sim G(\alpha_2, \beta_2)$	$\alpha_1 \leq \alpha_2$ $\beta_1 \leq \beta_2$	$\alpha_1 > \alpha_2$ $\beta_1 < \beta_2$ $\alpha_1 \beta_1 \leq \alpha_2 \beta_2$	$\alpha_1 > \alpha_2$ $\alpha_1 \beta_1 \leq \alpha_2 \beta_2$	$\alpha_1 \leq \alpha_2$ $\beta_1 \leq \beta_2$	$\alpha_1 \geq \alpha_2$
Weibull $X \sim W(\alpha_1, \beta_1)$ $Y \sim W(\alpha_2, \beta_2)$	$\beta_1 < \beta_2$ $\frac{\Gamma(1+1/\beta_1)}{\Gamma(1+1/\beta_2)} \leq \frac{\alpha_2}{\alpha_1}$	$\beta_1 > \beta_2$ $\frac{\Gamma(1+1/\beta_1)}{\Gamma(1+1/\beta_2)} \leq \frac{\alpha_2}{\alpha_1}$	$\beta_1 > \beta_2$ $\frac{\Gamma(1+1/\beta_1)}{\Gamma(1+1/\beta_2)} \leq \frac{\alpha_2}{\alpha_1}$	$\beta_1 > \beta_2$ $\frac{\Gamma(1+1/\beta_1)}{\Gamma(1+1/\beta_2)} \leq \frac{\alpha_2}{\alpha_1}$	$\beta_1 \geq \beta_2$
Pareto $X \sim P(a_1, k_1)$ $Y \sim P(a_2, k_2)$	$1 < a_1 < a_2$ $\frac{a_1(a_2-1)}{a_2(a_1-1)} \leq \frac{k_2}{k_1}$	$a_1, a_2 > 1$ $k_1 > k_2$ $\frac{a_1(a_2-1)}{a_2(a_1-1)} \leq \frac{k_2}{k_1}$	$a_1, a_2 > 1$ $k_1 > [=] k_2$ $\frac{a_1(a_2-1)}{a_2(a_1-1)} = [<] \frac{k_2}{k_1}$	$a_1 > a_2 > 1$ $a_1 k_2 > a_2 k_1$ $\frac{a_1(a_2-1)}{a_2(a_1-1)} \leq \frac{k_2}{k_1}$	$a_1 \geq a_2$

[82] and Taylor [83]. We want to mention that we have not considered the case of two parametric distributions belonging to different families. This topic have been considered by Lisek [82], Taylor [83], and Belzunce et al. [13] where some results can be found for the stochastic, increasing convex, increasing concave and star-shaped orders. Tables 2.3 and 2.4 gather the results obtained for the Davis and Gonvidarajulu families in this chapter. To finish, Table 2.5 provides several results for different families of discrete distributions. These results can be found in Ref. [84].

All of these results allow us to compare two random populations when their distributions are fitted by some of the parametric models considered in this section, whenever the parameters satisfy the appropriate restrictions.

Table 2.3 Comparison of some parametric continuous distributions in the hazard rate, the stochastic, the dispersive, and the star-shaped orders

Family		Criterion		
	$X \leq_{hr} Y$	$X \leq_{st} Y$	$X \leq_{disp} Y$	$X <_* Y$
Davies $X \sim D(\lambda_1, \theta_1, C_1)$ $Y \sim D(\lambda_2, \theta_2, C_2)$		$\lambda_1 > \lambda_2$ $\theta_1 < \theta_2$ $F^{-1}(p_0) \leq G^{-1}(p_0)$ $p_0 = \frac{\lambda_1 - \lambda_2}{\lambda_1 - \lambda_2 - \theta_1 + \theta_2}$		$\lambda_1 \leq \lambda_2$ $\theta_1 \leq \theta_2$
Govindarajulu $X \sim G(\beta_1, \sigma_1, \theta_1)$ $Y \sim G(\beta_2, \sigma_2, \theta_2)$	$\beta_2 \leq 1$ $\beta_1 \geq \beta_2$ $\theta_1 \leq \theta_2$ $\frac{(\beta_1+1)\beta_1}{(\beta_2+1)\beta_2} \leq \frac{\sigma_2}{\sigma_1}$	$\begin{cases} \beta_1 \geq \beta_2 \\ \theta_1 \leq \theta_2 \\ \frac{(\beta_1+1)\beta_1}{(\beta_2+1)\beta_2} \leq \frac{\sigma_2}{\sigma_1} \end{cases}$ or $\begin{cases} \beta_1 \leq \beta_2 \\ \theta_1 + \sigma_1 \leq \theta_2 + \sigma_2 \\ \frac{(\beta_1+1)\beta_1}{(\beta_2+1)\beta_2} \geq \frac{\sigma_2}{\sigma_1} \end{cases}$ or $\begin{cases} \beta_1 \leq \beta_2 \\ \frac{(\beta_1+1)\beta_1}{(\beta_2+1)\beta_2} \leq \frac{\sigma_2}{\sigma_1} \\ F^{-1}(p_0) \leq G^{-1}(p_0), \\ p_0 = \left(\frac{\sigma_2 \beta_2 (\beta_2+1)}{\sigma_1 \beta_1 (\beta_1+1)} \right)^{\frac{1}{\beta_1 - \beta_2}} \end{cases}$ or $\begin{cases} \beta_1 \geq \beta_2 \\ \theta_2 \geq \theta_1 \\ \theta_1 + \sigma_1 \leq \theta_2 + \sigma_2 \\ \frac{(\beta_1+1)\beta_1}{(\beta_2+1)\beta_2} \geq \frac{\sigma_2}{\sigma_1} \end{cases}$	$\beta_1 \geq \beta_2$ $\frac{(\beta_1+1)\beta_1}{(\beta_2+1)\beta_2} \geq \frac{\sigma_2}{\sigma_1}$	$\beta_1 \geq \beta_2$ $\frac{(\beta_1+1)\beta_1}{(\beta_2+1)\beta_2} \geq \frac{\theta_1+1}{\theta_2+1}$

Table 2.4 Comparison of some parametric continuous distributions in the ttt, the increasing convex, the excess wealth, and the expected proportional shortfall orders

Family	Criterion			
	$X \leq_{\mathrm{ttt}} Y$	$X \leq_{\mathrm{icx}} Y$	$X \leq_{\mathrm{ew}} Y$	$X \leq_{\mathrm{ps}} Y$
Davies $X \sim D(\lambda_1, \theta_1, C_1)$ $Y \sim D(\lambda_2, \theta_2, C_2)$	$\lambda_1 > [=]\lambda_2$ $1 > \theta_1 = [>]\theta_2$ $\frac{B(1+\lambda_1, 1-\theta_1)}{B(1+\lambda_2, 1-\theta_2)} \leq \frac{C_2}{C_1}$	$\lambda_1 < \lambda_2$ $\theta_1 \leq \theta_2 < 1$ $\frac{B(1+\lambda_1, 1-\theta_1)}{B(1+\lambda_2, 1-\theta_2)} \leq \frac{C_2}{C_1}$	$\lambda_1 < [=]\lambda_2$ $\theta_1 = [<]\theta_2 < 1$ $\frac{B(1+\lambda_1, 1-\theta_1)}{B(1+\lambda_2, 1-\theta_2)} \leq \frac{C_2}{C_1}$	$\lambda_1 \leq \lambda_2$ $\theta_1 \leq \theta_2$
Govindarajulu $X \sim G(\beta_1, \sigma_1, \theta_1)$ $Y \sim G(\beta_2, \sigma_2, \theta_2)$		$\beta_1 \leq \beta_2$ $\theta_1 \geq \theta_2$ $\frac{(\beta_1+1)\beta_1}{(\beta_2+1)\beta_2} \leq \frac{\sigma_2}{\sigma_1}$ $\theta_1 + \sigma_1 \leq \theta_2 + \sigma_2$ $\theta_1 + \frac{2\sigma_1}{\beta_1+2} \leq \theta_2 + \frac{2\sigma_2}{\beta_2+2}$	$\beta_1 < \beta_2$ $\theta_1 \geq \theta_2$ $\theta_1 + \frac{2\sigma_1}{\beta_1+2} \leq \theta_2 + \frac{2\sigma_2}{\beta_2+2}$	$\begin{cases} \theta_1 > 0 \\ \beta_1 \leq \beta_2 \\ \frac{\theta_2}{\theta+1} \leq \frac{\beta_1+2}{\beta_2+2} \end{cases}$ or $\begin{cases} \theta_1 = \theta_2 = 0 \\ \beta_1 \leq \beta_2 \end{cases}$

Table 2.5 Comparison of some parametric discrete distributions in the likelihood and the stochastic orders

Family	Criterion	
	$X \leq_{\mathrm{lr}} Y$	$X \leq_{\mathrm{st}} Y$
Binomial $X \sim B(n_1, p_1)$ $Y \sim B(n_2, p_2)$	$p_1 = 0$ or $\begin{cases} n_1 \leq 2 \\ \frac{n_1 p_1}{1-p_1} \leq \frac{n_2 p_2}{1-p_2} \end{cases}$	$n_1 \leq n_2$ $(1-p_1)^{n_1} \geq (1-p_2)^{n_2}$
Negative Binomial $X \sim BN(r_1, p_1)$ $Y \sim BN(r_2, p_2)$		$p_1 \leq p_2$ $p_1^{r_1} \geq p_2^{r_2}$
Poisson $X \sim P(\lambda_1)$ $Y \sim P(\lambda_2)$	$\lambda_1 \leq \lambda_2$	$\lambda_1 \leq \lambda_2$
Hypergeometric $X \sim H(N_1, m_1, n_1)$ $Y \sim H(N_2, m_2, n_1)$		$\begin{cases} N_1 \leq N_2 \\ p(x_0) \geq q(x_0) \\ p(x^0) \leq q(x^0) \end{cases}$ or $\begin{cases} \{n_1, m_1, n_2 + m_2 - N_2\} \cap \{n_2, m_2, n_1 + m_1 - N_1\} \neq \emptyset \\ p(x_0) \geq q(x_0) \\ p(x^0) \leq q(x^0) \end{cases}$ $x_0 = (n_1 + m_1 - N_1)_+ \wedge (n_2 + m_2 - N_2)_-$ $x^0 = (n_1 \wedge m_1) \vee (n_2 \wedge m_2)$

Besides the examples explicitly given in the chapter (and the references on the topic to which we have referred the reader), some of the conditions included in the different tables have been obtained directly from the previous ones and the relationships among the criteria. The following graph shows in which way they are related to each other.

$$X \leq_{\text{ttt}} Y \overset{l_X \leq l_Y}{\Rightarrow} X \leq_{\text{icv}} Y$$

$$\Uparrow$$

$$X \leq_{\text{lr}} Y \Rightarrow X \leq_{\text{hr}} Y \Rightarrow X \leq_{\text{st}} Y \overset{l_X \leq l_Y}{\Leftarrow} X \leq_{\text{disp}} Y$$

$$\Downarrow \qquad \Downarrow \qquad \Downarrow$$

$$X \leq_{\text{mrl}} Y \Rightarrow X \leq_{\text{icx}} Y \overset{l_X \leq l_Y}{\Leftarrow} X \leq_{\text{ew}} Y$$

$$X \leq_\star Y \Rightarrow X \leq_{\text{ps}} Y \Rightarrow X \leq_{\text{L}} Y$$

2.9.2 Comparison of coherent systems and distorted distributions

In reliability theory, one of the most important probabilistic models is the *coherent system*. The idea is to provide a mathematical description of a very common situation in reliability. In particular, a device which performs some function, for example, a car, which is made up of different components, motor, wheels, etc. Obviously, the operation of the device depends on the operation of the different components. Roughly speaking, we assume that the state of the system depends on each one of the components in a way such that, if one component fails, the system can fail but not the opposite, for more details see Ref. [10]. One of the main interest in this context is to obtain the survival function of the coherent system.

Let us consider a coherent system made up of n components with random lifetimes $X_1, \ldots, X_n$ and let us consider that the random lifetimes are independent. If we denote by T the random lifetime of such a coherent system, then we see that

$$P[T_{\mathbf{X}} > x] = h(\overline{F}_1(x), \ldots, \overline{F}_n(x)),$$

where $\overline{F}_i$ is the survival function of the random lifetime X_i, for all $i = 1, \ldots, n$ and the function $h : [0,1]^n \mapsto [0,1]$ is known as the *reliability function* of the system. The reliability function is strictly increasing and linear in p_i, for all $i = 1, \ldots, n$, and, in the particular case where the random lifetimes are equally distributed, then h can be expressed as a polynomial. Let us see some particular cases.

(a) *Series systems:* In a series system, the system fails when one of the components fails, that is $T = \min\{X_1,\ldots,X_n\}$ and the reliability function is given by

$$h(p_1,\ldots,p_n) = \prod_{i=1}^{n} p_i.$$

(b) *Parallel system:* In a parallel system, the system fails when every component has failed, that is $T = \max\{X_1,\ldots,X_n\}$, and the reliability function is given by

$$h(p_1,\ldots,p_n) = 1 - \prod_{i=1}^{n}(1 - p_i).$$

(c) *k-out-of-n system*: In a k-out-of-n system, the system works if, and only if, k or more of the components work. In the case where the components are equally distributed, the reliability function is given by

$$h(p) = \sum_{i=0}^{k-1}(1 - p)^i p^{n-i}.$$

By differentiation and assuming that the distribution function of every component is differentiable, we see that the density function of T is given by

$$f_T(t) = \sum_{i=1}^{n} f_i(t)\frac{\partial h}{\partial p_i}(\overline{F}_1(t),\ldots,\overline{F}_n(t)),$$

and the failure rate is given by

$$r_T(t) = \sum_{i=1}^{n} r_i(t)\left[\frac{p_i}{h\left(\overline{F}_1(t),\ldots,\overline{F}_n(t)\right)}\frac{\partial h}{\partial p_i}\left(\overline{F}_1(t),\ldots,\overline{F}_n(t)\right)\right], \quad (2.63)$$

where f_i and r_i denote the density and the hazard rate functions, respectively, of X_i, for all $i = 1,\ldots,n$.

Under the assumption of independent and identically distributed (i.i.d.) random lifetimes, it is very easy to provide results for the comparison of two coherent systems. For example, let us consider two coherent systems T_1 and T_2 with reliability functions h_1 and h_2, which are made up of two sets of independent and identically distributed random lifetimes $X_1,\ldots,X_n$ and $Y_1,\ldots,Y_n$ with common survival functions $\overline{F}$ and $\overline{G}$, respectively. Let us assume that $\overline{F}(x) \leq \overline{G}(x)$, for all $x \in \mathbb{R}$, that is, the stochastic order holds among any pair of the components, and let us assume that $h_1(p) \leq h_2(p)$, for all $p \in [0, 1]$, then $T_1 \leq_{\text{st}} T_2$. Similar results can be given for the likelihood ratio, hazard rate, and dispersive orders [85–87].

The previous result for coherent systems with i.i.d. components was extended to the independent, but not necessarily identically distributed (i.n.i.d.) distributions, by Belzunce et al. [88]. For instance, let us consider two coherent systems T_1 and T_2 as above where the components are no identically distributed, their survival functions are given by $\overline{F}_i$ and $\overline{G}_i$, and the hazard rates functions by r_i and s_i, respectively, for all $i = 1, \ldots, n$. Let us assume that $X_i \leq_{\text{st}} Y_i$, for all $i = 1, \ldots, n$,

$$\min_{i=1,\ldots,n} \{r_i(x)\} \geq \max_{i=1,\ldots,n} \{s_i(x)\}, \quad \text{for all } x \in \mathbb{R}^+, \tag{2.64}$$

$$\sum_{i=1}^{n} \frac{p_i}{h_1(p_1,\ldots,p_n)} \frac{\partial h_1}{\partial p_i}(p_1,\ldots,p_n) \tag{2.65}$$

$$\leq \sum_{i=1}^{n} \frac{p_i}{h_2(p_1,\ldots,p_n)} \frac{\partial h_2}{\partial p_i}(p_1,\ldots,p_n),$$

for all $(p_1, \ldots, p_n) \in (0, 1)^n$, and

$$\sum_{i=1}^{n} \frac{p_i}{h_1(p_1,\ldots,p_n)} \frac{\partial h_1}{\partial p_i}(p_1,\ldots,p_n) \text{ is decreasing in } p_j, \tag{2.66}$$

for all $j = 1, \ldots, n$. Recalling (2.63), we have the following inequalities:

$$r_{T_1}(x) = \sum_{i=1}^{n} r_i(x) \left[\frac{p_i}{h_1\left(\overline{F}_1(x),\ldots,\overline{F}_n(x)\right)} \frac{\partial h_1}{\partial p_i} \left(\overline{F}_1(x),\ldots,\overline{F}_n(x)\right) \right]$$

$$\geq \min_{i=1,\ldots,n} \{r_i(x)\} \sum_{i=1}^{n} \left[\frac{p_i}{h_1\left(\overline{F}_1(x),\ldots,\overline{F}_n(x)\right)} \frac{\partial h_1}{\partial p_i} \left(\overline{F}_1(x),\ldots,\overline{F}_n(x)\right) \right]$$

$$\geq \max_{i=1,\ldots,n} \{s_i(x)\} \sum_{i=1}^{n} \left[\frac{p_i}{h_1\left(\overline{G}_1(x),\ldots,\overline{G}_n(x)\right)} \frac{\partial h_1}{\partial p_i} \left(\overline{G}_1(x),\ldots,\overline{G}_n(x)\right) \right]$$

$$\geq \sum_{i=1}^{n} s_i(t) \left[\frac{p_i}{h_2\left(\overline{G}_1(x),\ldots,\overline{G}_n(x)\right)} \frac{\partial h_2}{\partial p_i} \left(\overline{G}_1(x),\ldots,\overline{G}_n(x)\right) \right] = r_{T_2}(x),$$

for all $x \in \mathbb{R}^+$, where the second inequality follows from (2.64) and (2.66) and the third one by (2.65). Therefore, $T_1 \leq_{\text{hr}} T_2$.

In Ref. [88] the reader can find many results for the stochastic, hazard rate, reversed hazard rate, and likelihood ratio orders for the i.n.i.d. case. Some recent papers for the case of coherent systems with not necessarily independent components can be seen in Refs. [89–91].

Previous results allow us to give another result for the particular case of k-out-of-n systems. It is well known that (2.66) holds for k-out-of-n systems [92] and, therefore, for two k-out-of-n systems made up of two sets of i.n.i.d. non-negative random lifetimes $X_1, \ldots, X_n$ and $X_1, \ldots, Y_n$, where $X_i \leq_{\mathrm{st}} Y_i$, for all $i = 1, \ldots, n$, we see that the corresponding k-out-of-n systems are ordered in the hazard rate order, if (2.64) holds. Recall that k-out-of-n systems can be seen as the order statistics from the random lifetimes $X_1, \ldots, X_n$. In particular, the random lifetime of a k-out-of-n system made up of components with random lifetimes $X_1, \ldots, X_n$ is the order statistic $X_{n-k+1:n}$. Therefore, the previous comparison of k-out-of-n systems can be considered as results for the comparison of order statistics. There are a great many papers devoted to the comparison of order statistics; we refer the reader to Refs. [93, 94] for reviews on the topic.

The survival function of a coherent system in the i.i.d. case can be seen as a particular case of *distorted distributions*. A function $h : [0, 1] \mapsto [0, 1]$ is called a distortion if h is increasing, left continuous and such that $h(0) = 0$ and $h(1) = 1$. Given a survival function $\overline{F}$, the composition $h(\overline{F})$, where h is a distortion function, can be considered as a survival function as well. Distorted distributions has great interest in actuarial sciences and have been applied to a variety of insurance problems, for example, to the determination of insurance premiums and risk measures (for further details, see Ref. [95, 96] and Section 2.6 in [6], and the references therein). In reliability theory, there are also applications of distorted distributions. In particular, the survival function of a coherent system with i.i.d. components is the distortion of the common survival function of the components through the reliability function. Recently, Navarro et al. [97] have noticed that the survival function of a coherent system based on possibly dependent but identically distributed components with random lifetimes $X_1, \ldots, X_n$ and common survival function $\overline{F}$, can be expressed as a distortion of the survival function $h(\overline{F})$, where the distortion function h depends only on the structure of the system and the survival copula of $(X_1, \ldots, X_n)$.

It is also possible to provide distortions of multivariate distributions. Next, we discuss external and global distorsions introduced by di Bernardino and Rullière [98].

An *external* *distortion* of a n dimensional survival function is an increasing function $h : [0, 1] \mapsto [0, 1]$, where $h(0) = 0$, $h(1) = 1$ and such that $h(\overline{F})$ is a multivariate survival function for any n dimensional survival function $\overline{F}$. Under some regularity conditions on the survival functions, di Bernardino and Rullière [98] prove that if h is continuous on $[0, 1]$ and has non-negative derivatives of orders up to, and including n, on $(0, 1)$, then h is an external distortion. Given an n dimensional survival function $\overline{F}$, the external distortion of $\overline{F}$ through h is given by

$$h(\overline{F}(x_1, \ldots, x_n)) = h(C(\overline{F}_1(x_1), \ldots, \overline{F}_n(x_n))),$$

where C is the survival copula (which plays the same role that the copula for the multivariate distribution function, but for the multivariate survival function) associated to $\overline{F}$ and $\overline{F}_i$, for all $i = 1, \ldots, n$, denote the survival functions of the marginals. We observe that the survival function of a coherent system in the i.n.i.d. case can be considered as an example of an external distortion.

In a *global distortion*, there is an external distortion, h, as above, and n continuous distortion functions $h_i : [0, 1] \mapsto [0, 1]$. Under previous notation, given an n dimensional survival function $\overline{F}$, the global distortion of $\overline{F}$ through $h, h_1, \ldots, h_n$, is given by

$$h(C(h_1(\overline{F}_1(x_1)), \ldots, h_n(\overline{F}_n(x_n)))).$$

In both cases, we have a transformation through a function $H : [0, 1]^n \mapsto [0, 1]$ of the marginal survival functions $\overline{F}_i$, for all $i = 1, \ldots, n$.

From the previous discussion, we observe that we can derive results for the comparison of univariate or multivariate distortions from previous results in the literature on the comparison of coherent systems in the i.i.d. and i.n.i.d. cases, respectively.

2.9.3 Comparison of individual and collective risks and shock models

In actuarial science and insurance companies, there are two situations which are particular cases of the convolution of random variables, named individual and collective risks. In the individual risk model, we assume that the portfolio of an insurer consists of n policies. Denoting by X_i the payment on policy i, for all $i = 1, \ldots, n$, then the total claim of the portfolio is given by

$$S = X_1 + \cdots + X_n.$$

In the collective risk model, we consider a portfolio that produces a claim at random points in time. In particular, we see that the total claim of the portfolio is given by

$$S_N = X_1 + \cdots + X_N,$$

where X_i is the claim of the policy i, and N denotes the random number of claims.

Usual distributional assumptions in both cases are that the claims X_i are independent and identically distributed and, in the collective risk model, that the number of claims N and X_i are independent. The claims X_i are assumed to be a mixture of a degenerate random variable at point $x = 0$ and a positive continuous random variable where the mixing distribution is a Bernoulli distribution. In this case, when there is a loss, the loss amount is a continuous random variable U_i, and with probability p there is no claim. Common parametric models for the random loss are the exponential, gamma, Pareto, and lognormal distributions, whereas common parametric models for N are Poisson, binomial, negative binomial, and logarithmic distributions.

Next, we describe some results for the comparison of individual and risks models.

Let us consider two individual risk models for two sets of random claims $X_1, \ldots, X_n$ and $Y_1, \ldots, Y_n$. Let us assume that there is no loss, with probabilities p_1 and p_2, and there are losses U_i and V_i, with probabilities $1 - p_1$ and $1 - p_2$, respectively. If we assume that $p_1 \geq p_2$ and $U_i \leq_{st} V_i$, for all $i = 1, \ldots, n$, then, from Theorem 2.2.8, we see that $X_i \leq_{st} Y_i$, for all $i = 1, \ldots, n$. Therefore, from Theorem 2.2.9, we see that

$$S_1 = \sum_{i=1}^{n} X_i \leq_{st} S_2 = \sum_{i=1}^{n} Y_i,$$

and, consequently, $\mathrm{VaR}[S_1; p] \leq \mathrm{VaR}[S_2; p]$, for all $p \in (0, 1)$.

If we have the weaker assumption $U_i \leq_{icx} V_i$, for all $i = 1, \ldots, n$, then from Theorems 2.3.12 and 2.3.13, we have

$$S_1 = \sum_{i=1}^{n} X_i \leq_{icx} S_2 = \sum_{i=1}^{n} Y_i,$$

and, therefore, $\mathrm{TVaR}[S_1; p] \leq \mathrm{TVaR}[S_2; p]$, for all $p \in (0, 1)$.

For example, let us assume that the losses are distributed according to gamma distributions, that is, $U_i \sim G(\alpha_1, \beta_1)$ and $V_i \sim G(\alpha_2, \beta_2)$, for $i = 1, 2$. From Table 2.1, if $\alpha_1 \leq \alpha_2$ and $\beta_1 \leq \beta_2$, then $U_i \leq_{st} V_i$, for $i = 1, 2$. Therefore, the corresponding individual risk models are ordered in the stochastic order. If the conditions on the parameters are $\alpha_1 > \alpha_2$ and $\alpha_1 \beta_1 < \alpha_2 \beta_2$, then $U_i \leq_{icx} V_i$, for $i = 1, 2$, from Table 2.2 and, consequently, the corresponding individual risk models are ordered in the icx order.

We want to point out that the comparison of finite convolutions for some specific parametric families of distributions have received a great attention in the literature. A review of this topic can be found in Ref. [99].

These comparisons of individual risk models can be extended to the comparison of collective risk models. Let us consider two individual risk models for two sets of random claims $\{X_n\}_{n \in \mathbb{N}}$ and $\{Y_n\}_{n \in \mathbb{N}}$, with random number of claims N_1 and N_2, respectively. According to the previous discussion for the individual risk models, if we also consider that $N_1 \leq_{st} N_2$ or $N_1 \leq_{icx} N_2$, then we can provide comparisons in the st or icx orders for the corresponding collective risks models. The results follow by considering a random sum $\sum_{i=1}^{N} X_i$ as a mixture of the family $\{\sum_{i=1}^{n} X_i | n \in \mathbb{N}\}$ with the discrete random variable N. By Theorems 2.2.8 and 2.3.12 we can provide comparisons of collective risk models. For more details and additional results, see Ref. [2].

Finally, another application to the comparison of *shock models* is shown; these models are widely used in reliability to describe the following situation. Let us consider a device that is subjected to shocks occurring randomly in time. The device is assumed to has a probability of surviving the first k shocks, for all $k \in \mathbb{N}$, denoted by $\overline{P}_k$. Usually, the sequence $\{\overline{P}_k\}_{k \in \mathbb{N}}$ is assumed to be decreasing and $\lim_{k \to +\infty} \overline{P}_k = 0$, therefore, it can be considered as the survival function of a discrete non-negative random variable N. If we denote by $\{X_n\}_{n \in \mathbb{N}}$ the sequence of interarrival times between two consecutive shocks, and assuming that N is independent of the interarrival times, then the random lifetime of the device until a fatal shock occurs is given by $\sum_{i=1}^{N} X_i$. Therefore, previous results for the comparison of random convolutions can be applied when the interarrival times are independent. This is the case when shocks occurs according to a homogeneous Poisson process, where the interarrival times are independent and exponentially distributed, or the more general case, when shocks occurs according to a renewal process (independent and identically distributed interarrivals).

Two other cases not covered by the previous discussion, in which interarrivals are not independent, are the cases where shocks occurs according to a non-homogeneous Poisson process or to a non-homogeneous pure birth process. In this cases, even in the more general case of non-independent random lifetimes, is possible to provide results as follows. Let us consider two shock models with interarrival sequences $\{X_n\}_{n\in\mathbb{N}}$ and $\{Y_n\}_{n\in\mathbb{N}}$ and discrete random variables N_1 and N_2 describing the survival probabilities of the number of shocks, respectively. Let us assume that

$$\sum_{i=1}^{n} X_i \leq_{st} \sum_{i=1}^{n} Y_i, \quad \text{for all } n \in \mathbb{N}.$$

If $N_1 \leq_{st} N_2$, then by Theorem 2.2.8, we see that

$$\sum_{i=1}^{N_1} X_i \leq_{st} \sum_{i=1}^{N_2} Y_i.$$

In the particular cases where shocks occurs according to a nonhomogeneous Poisson process or to a nonhomogeneous pure birth process, it is possible to provide results for $\sum_{i=1}^{n} X_i \leq_{st} \sum_{i=1}^{n} Y_i$, where the random variables $\sum_{i=1}^{n} X_i$ and $\sum_{i=1}^{n} Y_i$ are known as the epoch times. Results in this direction can be seen in Ref. [100]. In any case, some additional results for the comparison of epoch times of previous processes will be given in Section 3.7. Some other references on the topic of shock models are [101–107].

2.10 SUMMARY

This chapter gives an overview of some of the main criteria for the several stochastic comparisons of two random variables. Among the great number of criteria in the literature, we have considered in this chapter those which have a direct application in several context like reliability, risk theory, and economics. Furthermore, these criteria are among those which receive a greater attention and they can be considered the most prominent ones. As a general scheme for the introduction of these criteria, we have provided the main characterizations, some sets of sufficient conditions, preservation results and graphical procedures. In particular, one of the main contributions of this chapter is to provide several results on stochastic comparisons of parametric families of univariate distributions, filling a gap in the previous literature. We have provided a review on the topic, as much complete as possible, collecting in Tables 2.1–2.5 the results obtained in the chapter, jointly with previous results given by Taylor [83], Lisek [82], and Klenke and Mattner [84]. The idea is to facilitate the reader the search of results. Chapter 2 finishes with an overview on applications in some stochastic models of interest in reliability and risks, like coherent systems distorted distributions, individual and collective risk and shock models. In conclusion, Chapter 2 is intended as a suitable guide for anyone interested in initiating themselves in the theory of univariate stochastic orders, as well as a basic reference on the comparison of parametric models.

CHAPTER 3

Multivariate stochastic orders

3.1 INTRODUCTION

A natural question that arises when dealing with stochastic orders is how to extend a univariate stochastic order to the multivariate case—that is, how to compare two random vectors with the same dimension in an analogous sense to those in Chapter 2. These generalizations can be made by means of the different characterizations taken into account in the univariate case. Some of these characterizations are stated from a mathematical point of view and others are motivated from an applied one. Here, some extensions of the univariate stochastic orders considered in the previous chapter are given. In particular, we recall those that we consider to be the most interesting and/or useful from an applied one. We refer the reader to Refs. [1–3] for a complete review on multivariate stochastic orders. In this chapter, only a few examples of ordered parametric models are provided. This is because this topic has not been covered extensively in the literature. We do, however, refer the reader to Refs. [108–110] for several results in this direction. This chapter finishes with the description of some applications on conditionally independent random variables and several results for the comparison of ordered data.

3.2 THE MULTIVARIATE USUAL STOCHASTIC ORDER

First, a multivariate extension of the usual stochastic order is considered. The following definition is based on some of the characterizations provided for the univariate stochastic order (see, in particular, Theorem 2.2.5 (ii) and (iii)).

Definition 3.2.1. Given two random vectors $\mathbf{X} = (X_1, \dots, X_n)$ and $\mathbf{Y} = (Y_1, \dots, Y_n)$, we say that $\mathbf{X}$ is smaller than $\mathbf{Y}$ in the *multivariate stochastic order*, denoted by $\mathbf{X} \leq_{st} \mathbf{Y}$, if

$$E[\phi(\mathbf{X})] \leq E[\phi(\mathbf{Y})],$$

An Introduction to Stochastic Orders. http://dx.doi.org/10.1016/B978-0-12-803768-3.00003-X

for all increasing function $\phi : \mathbb{R}^n \mapsto \mathbb{R}$, such that the previous expectations exist, or, equivalently, if

$$\phi(\mathbf{X}) \leq_{\text{st}} \phi(\mathbf{Y}),$$

for all increasing function $\phi : \mathbb{R}^n \mapsto \mathbb{R}$.

The equivalence among the two previous conditions follows from Theorem 2.2.5(ii) and (iii). It is also possible to characterize the multivariate stochastic order by the construction of proper random vectors on the same probability space, analogously to Theorem 2.2.5(i). The proof is omitted and can be seen in Ref. [111].

Theorem 3.2.2. *Let* $\mathbf{X} = (X_1, \ldots, X_n)$ *and* $\mathbf{Y} = (Y_1, \ldots, Y_n)$ *be two random vectors. Then,* $\mathbf{X} \leq_{\text{st}} \mathbf{Y}$ *if, and only if, there exist two random vectors* $\widehat{\mathbf{X}}$ *and* $\widehat{\mathbf{Y}}$ *defined on the same probability space, such that* $\widehat{\mathbf{X}} =_{\text{st}} \mathbf{X}$, $\widehat{\mathbf{Y}} =_{\text{st}} \mathbf{Y}$ *and* $P[\widehat{\mathbf{X}} \leq \widehat{\mathbf{Y}}] = 1$.

Recall that an important result in the univariate case is the preservation of the stochastic order under convolution (see Theorem 2.2.9). Let us observe that, from Definition 3.2.1, if $(X_1, \ldots, X_n) \leq_{\text{st}} (Y_1, \ldots, Y_n)$, then

$$\sum_{i=1}^{n} X_i \leq_{\text{st}} \sum_{i=1}^{n} Y_i. \tag{3.1}$$

In fact, given a set of independent random variables $X_1, \ldots, X_n$ and another set of independent random variables $Y_1, \ldots, Y_n$, such that $X_i \leq_{\text{st}} Y_i$, for all $i = 1, \ldots, n$, we see that $(X_1, \ldots, X_n) \leq_{\text{st}} (Y_1, \ldots, Y_n)$ and, therefore, (3.1) holds.

Hence, in order to provide general results for the stochastic order under convolutions of not necessarily independent random variables, it is enough to prove the multivariate stochastic order among the corresponding random vectors. Next, several results which provide sufficient conditions for the multivariate stochastic order based on the univariate stochastic order are given.

Theorem 3.2.3. *Let* $\mathbf{X} = (X_1, \ldots, X_n)$ *and* $\mathbf{Y} = (Y_1, \ldots, Y_n)$ *be two random vectors. If*

$$X_1 \leq_{\text{st}} Y_1, \tag{3.2}$$

and

$$[X_i | X_1 = x_1, \ldots, X_{i-1} = x_{i-1}] \leq_{\text{st}} [Y_i | Y_1 = y_1, \ldots, Y_{i-1} = y_{i-1}], \tag{3.3}$$

for all $x_j \leq y_j$, $j = 1, \ldots, i - 1$, *in the support of* $X_1, \ldots, X_{i-1}$ *and* $Y_1, \ldots, Y_{i-1}$, *respectively, for all* $i = 2, \ldots, n$, *then*

$$\mathbf{X} \leq_{st} \mathbf{Y}.$$

Proof. As we see next, the proof follows from Theorem 3.2.2. Recalling the standard construction (see Section 1.3), let us consider the random vectors $\hat{\mathbf{X}} = \mathbf{Q}_{\mathbf{X}}(U_1, \ldots, U_n)$ and $\hat{\mathbf{Y}} = \mathbf{Q}_{\mathbf{Y}}(U_1, \ldots, U_n)$, where $U_1, \ldots, U_n$ are independent and uniformly distributed random variables on the interval $(0, 1)$. Obviously, $\hat{\mathbf{X}}$ and $\hat{\mathbf{Y}}$ are defined on the same probability space. From (1.12), we see that $\hat{\mathbf{X}} =_{st} \mathbf{X}$ and $\hat{\mathbf{Y}} =_{st} \mathbf{Y}$. Given $(u_1, \ldots, u_n) \in (0, 1)^n$, from (3.2), (3.3) and Theorem 2.2.3, it is easy to see that $\mathbf{Q}_{\mathbf{X}}(u_1, \ldots, u_n) \leq \mathbf{Q}_{\mathbf{Y}}(u_1, \ldots, u_n)$, therefore $P[\hat{\mathbf{X}} \leq \hat{\mathbf{Y}}]$ and the result follows.   □

Assuming the CIS dependence property (see Section 1.3) for one of the random vectors, it is possible to weaken the condition (3.3) in the previous theorem. In particular, we have the following theorem.

Theorem 3.2.4. *Let* $\mathbf{X} = (X_1, \ldots, X_n)$ *and* $\mathbf{Y} = (Y_1, \ldots, Y_n)$ *be two random vectors. Let us assume that* $\mathbf{X}$ *or* $\mathbf{Y}$ *or both are CIS and*

$$X_1 \leq_{st} Y_1,$$

and

$$[X_i | X_1 = x_1, \ldots, X_{i-1} = x_{i-1}] \leq_{st} [Y_i | Y_1 = x_1, \ldots, Y_{i-1} = x_{i-1}], \quad (3.4)$$

for all $x_1, \ldots, x_{i-1}$ *in the support of* $X_1, \ldots, X_{i-1}$ *and* $Y_1, \ldots, Y_{i-1}$, *respectively, for all* $i = 2, \ldots, n$, *then*

$$\mathbf{X} \leq_{st} \mathbf{Y}.$$

Proof. Notice that if $\mathbf{X}$ or $\mathbf{Y}$ or both are CIS, then (3.4) implies (3.3). The result follows from Theorem 3.2.3.   □

It is also possible to weaken the CIS property by the PA property (see Section 1.3). Next, an additional result with sufficient conditions for the multivariate stochastic order is provided.

Theorem 3.2.5. *Let* $\mathbf{X} = (X_1, \ldots, X_n)$ *and* $\mathbf{Y} = (Y_1, \ldots, Y_n)$ *be two continuous random vectors with common support S and joint density functions f and g, respectively. Let us assume that* $\mathbf{X}$ *or* $\mathbf{Y}$ *or both are PA. If* $g(\mathbf{x})/f(\mathbf{x})$ *is increasing in S, then*

$$\mathbf{X} \leq_{st} \mathbf{Y}.$$

Proof. Let $\phi : \mathbb{R}^n \mapsto \mathbb{R}$ be an increasing function, such that $E[\phi(\mathbf{X})]$ and $E[\phi(\mathbf{Y})]$ exist. Let us assume that $\mathbf{X}$ is PA (the proof when $\mathbf{Y}$ is PA is similar).

Then,

$$E[\phi(\mathbf{Y})] = \int \phi(\mathbf{x})g(\mathbf{x})\, d\mathbf{x} = \int \phi(\mathbf{x})\frac{g(\mathbf{x})}{f(\mathbf{x})}f(\mathbf{x})\, d\mathbf{x}$$

$$\geq \int \phi(\mathbf{x})f(\mathbf{x})\, d\mathbf{x} \int \frac{g(\mathbf{x})}{f(\mathbf{x})}f(\mathbf{x})\, d\mathbf{x} = E[\phi(\mathbf{X})],$$

where the inequality follows from the PA property, since ϕ and g/f are increasing functions. □

To conclude with the results about sufficient conditions for the multivariate stochastic order, the case of two random vectors with the same copula is considered. As we shall see later, the following theorem is very useful to provide results for the comparison of multivariate parametric distributions.

Theorem 3.2.6. *Let* $\mathbf{X} = (X_1, \ldots, X_n)$ *and* $\mathbf{Y} = (Y_1, \ldots, Y_n)$ *be two random vectors with a common copula* C. *If* $X_i \leq_{st} Y_i$, *for all* $i = 1, \ldots, n$, *then*

$$\mathbf{X} \leq_{st} \mathbf{Y}.$$

Proof. Let us denote by $\mathbf{U} = (U_1, \ldots, U_n)$ the random vector with copula C and let $\hat{\mathbf{X}} = (F_1^{-1}(U_1), \ldots, F_n^{-1}(U_n))$ and $\hat{\mathbf{Y}} = (G_1^{-1}(U_1), \ldots, G_n^{-1}(U_n))$, which are defined on the same probability space, where F_i and G_i denote the distribution functions of X_i and Y_i, respectively, for all $i = 1, \ldots, n$. In this case, from the properties of the copula, we see that $\hat{\mathbf{X}} =_{st} \mathbf{X}$ and $\hat{\mathbf{Y}} =_{st} \mathbf{Y}$. Now, from the condition $X_i \leq_{st} Y_i$, for all $i = 1, \ldots, n$, it is clear that

$$(F_1^{-1}(p_1), \ldots, F_n^{-1}(p_n)) \leq (G_1^{-1}(p_1), \ldots, G_n^{-1}(p_n)),$$

for all $(p_1, \ldots, p_n) \in (0, 1)^n$, and $P[\hat{\mathbf{X}} \leq \hat{\mathbf{Y}}]$ holds. The result follows from Theorem 3.2.2. □

From this result, a lot of examples for the comparison in the multivariate stochastic order can be provided. We have just to consider random vectors with the same copula, whatever it would be, and to consider univariate marginal distributions ordered in the stochastic order. Examples of univariate distributions ordered in the stochastic order can be found in Tables 2.1 and 2.3. Let us see an application to the case of multivariate normal distributions.

Example 3.2.7. Let $\mathbf{X} \sim N_n(\boldsymbol{\mu}_1, \boldsymbol{\Sigma}_1)$ and $\mathbf{Y} \sim N_n(\boldsymbol{\mu}_2, \boldsymbol{\Sigma}_2)$. As mentioned in Section 1.3, if they have the same correlation matrix, then $\mathbf{X}$ and $\mathbf{Y}$ share a common copula and, therefore, if $X_i \leq_{\mathrm{st}} Y_i$, for all $i = 1, \ldots, n$, then $\mathbf{X} \leq_{\mathrm{st}} \mathbf{Y}$. Since $X_i \sim N(\mu_{1i}, \sigma_{1i}^2)$ and $Y_i \sim N(\mu_{2i}, \sigma_{2i}^2)$, we see that, from Table 2.1, if $\mu_{1i} \leq \mu_{2i}$ and $\sigma_{1i}^2 = \sigma_{2i}^2$, then $X_i \leq_{\mathrm{st}} Y_i$, for all $i = 1, \ldots, n$. Notice that if $\mathbf{X}$ and $\mathbf{Y}$ share the same correlation copula and $\sigma_{1i}^2 = \sigma_{2i}^2$, for all $i = 1, \ldots, n$, equivalently, we have $\boldsymbol{\Sigma}_1 = \boldsymbol{\Sigma}_2$. To sum up, if $\boldsymbol{\mu}_1 \leq \boldsymbol{\mu}_2$ and $\boldsymbol{\Sigma}_1 = \boldsymbol{\Sigma}_2$, then $\mathbf{X} \leq_{\mathrm{st}} \mathbf{Y}$.

Next, some additional results on the preservation of the multivariate stochastic order are provided. First, a result on the preservation under increasing transformations is given.

Theorem 3.2.8. *Let* $\mathbf{X} = (X_1, \ldots, X_n)$ *and* $\mathbf{Y} = (Y_1, \ldots, Y_n)$ *be two random vectors. If* $\mathbf{X} \leq_{\mathrm{st}} \mathbf{Y}$*, then*

$$\phi(\mathbf{X}) \leq_{\mathrm{st}} \phi(\mathbf{Y}),$$

for all increasing function $\phi : \mathbb{R}^n \mapsto \mathbb{R}^k$.

Proof. The proof follows by the preservation of the univariate stochastic order under increasing transformations (see Theorem 2.2.5(ii)). $\square$

An obvious application of the previous result is the preservation of the multivariate stochastic order under marginalization.

Theorem 3.2.9. *Let* $\mathbf{X} = (X_1, \ldots, X_n)$ *and* $\mathbf{Y} = (Y_1, \ldots, Y_n)$ *be two random vectors. If* $\mathbf{X} \leq_{\mathrm{st}} \mathbf{Y}$*, then*

$$\mathbf{X}_I \leq_{\mathrm{st}} \mathbf{Y}_I, \quad \text{for all } I \subseteq \{1, \ldots, n\}.$$

Finally, some results on the preservation under mixtures are provided. One of the main applications of this kind of results is to study the case of conditionally independent random vectors, which is described next.

Let $(X_1, \ldots, X_n, \boldsymbol{\Theta}_1)$ and $(Y_1, \ldots, Y_n, \boldsymbol{\Theta}_2)$ be two random vectors, where $\boldsymbol{\Theta}_1$ and $\boldsymbol{\Theta}_2$ are m-dimensional random vectors with common support S. Generally, unless stated otherwise, we shall assume that $X_1, \ldots, X_n$ and $Y_1, \ldots, Y_n$ are independent random variables given $\boldsymbol{\Theta}_1 = \boldsymbol{\theta}$ and $\boldsymbol{\Theta}_2 = \boldsymbol{\theta}$, for any value of $\boldsymbol{\theta}$, respectively. Next, sufficient conditions for the multivariate stochastic order among $(X_1, \ldots, X_n)$ and $(Y_1, \ldots, Y_n)$ are described (for further details, see Ref. [112]).

Theorem 3.2.10. *According to the previous notation, assume that*

(i) $[X_i|\Theta_1 = \theta]$ *or* $[Y_i|\Theta_2 = \theta]$ *or both are increasing in the stochastic order in* θ, *for all* $i = 1, \ldots, n$,
(ii) $[X_i|\Theta_1 = \theta] \leq_{st} [Y_i|\Theta_2 = \theta]$, *for any* $\theta \in S$ *and for all* $i = 1, \ldots, n$, *and*
(iii) $\Theta_1 \leq_{st} \Theta_2$.

Then,

$$(X_1, \ldots, X_n) \leq_{st} (Y_1, \ldots, Y_n).$$

Proof. Let $\phi : \mathbb{R}^n \mapsto \mathbb{R}$ be an increasing function and let us assume that (i) holds for $X_i(\theta) = [X_i|\Theta_1 = \theta]$ (the proof in the other case is similar). Then, it is easy to see that

$$E[\phi(X_1(\theta), \ldots, X_n(\theta))] \text{ is increasing in } \theta. \tag{3.5}$$

As mentioned at the beginning of this section, (ii) is equivalent to

$$(X_1(\theta), \ldots, X_n(\theta)) \leq_{st} (Y_1(\theta), \ldots, Y_n(\theta)), \quad \text{for all } \theta \in S. \tag{3.6}$$

Now, denoting by H_1 and H_2 the joint distribution function of Θ_1 and Θ_2, respectively, we have the following chain of inequalities (we assume that the conditions of Fubini's theorem hold):

$$E[\phi(X_1, \ldots, X_n)] = \int_{\mathbb{R}^m} E[\phi(X_1(\theta), \ldots, X_n(\theta))] \, dH_1(\theta)$$

$$\geq \int_{\mathbb{R}^m} E[\phi(X_1(\theta), \ldots, X_n(\theta))] \, dH_2(\theta)$$

$$\geq \int_{\mathbb{R}^m} E[\phi(Y_1(\theta), \ldots, Y_n(\theta))] \, dH_2(\theta)$$

$$= E[\phi(Y_1, \ldots, Y_n)],$$

where the first inequality follows from (iii) and (3.5), and the second inequality, from (3.6). Therefore, $(X_1, \ldots, X_n) \leq_{st} (Y_1, \ldots, Y_n)$. □

The previous theorem is a particular case of the following one.

Theorem 3.2.11. *Let* $(X_1, \ldots, X_n, \Theta_1)$ *and* $(Y_1, \ldots, Y_n, \Theta_2)$ *be two random vectors where* $[X_1, \ldots, X_n|\Theta_1 = \theta]$ *are not necessarily independent, for all* $\theta \in S$ *(and analogously for* $[Y_1, \ldots, Y_n|\Theta_2 = \theta)]$). *If* *(i)* $[X_1, \ldots, X_n | \Theta_1 = \boldsymbol{\theta}]$ *or* $[Y_1, \ldots, Y_n | \Theta_2 = \boldsymbol{\theta})]$ *or both are increasing in $\boldsymbol{\theta}$ in the multivariate stochastic order,*

(ii) $[X_1, \ldots, X_n | \Theta_1 = \boldsymbol{\theta})] \leq_{\mathrm{st}} [Y_1, \ldots, Y_n | \Theta_2 = \boldsymbol{\theta}])$*, for any $\boldsymbol{\theta} \in S$, and*

(iii) $\Theta_1 \leq_{\mathrm{st}} \Theta_2$,

then,

$$(X_1, \ldots, X_n) \leq_{\mathrm{st}} (Y_1, \ldots, Y_n).$$

Proof. The proof follows under similar steps to those of Theorem 3.2.10. $\qquad\square$

In the conditionally independent case, (i) and (ii) in Theorem 3.2.10 are equivalent to (i) and (ii) in Theorem 3.2.11, respectively.

As we have seen, the multivariate stochastic order is defined in terms of increasing functions. Now, it is natural to wonder about the comparison of joint survival functions like in the univariate case. The following discussion is related to this question.

Let us consider an upper set $U \subseteq \mathbb{R}^n$ and the increasing function $I_U(\mathbf{x})$. If $\mathbf{X} \leq_{\mathrm{st}} \mathbf{Y}$, then $E[I_U(\mathbf{X})] \leq E[I_U(\mathbf{X})]$. Furthermore, if the previous inequality holds for any upper set $U \subseteq \mathbb{R}^n$, then $\mathbf{X} \leq_{\mathrm{st}} \mathbf{Y}$ [2].

Particularly, given the upper set $U = \prod_{i=1}^n (x_i, +\infty)$, for $(x_1, \ldots, x_n) \in \mathbb{R}^n$, if $\mathbf{X} \leq_{\mathrm{st}} \mathbf{Y}$, then

$$P[X_1 > x_1, \ldots, X_n > x_n] = E[I_U(\mathbf{X})] \tag{3.7}$$

$$\leq E[I_U(\mathbf{Y})] = P[Y_1 > x_1, \ldots, Y_n > x_n],$$

for all $(x_1, \ldots, x_n) \in \mathbb{R}^n$. However, the inverse implication is not true in general. In fact, (3.7) is used to define a multivariate stochastic order known as the upper orthant order. For further details on this criterion and related orders, we refer the reader to Ref. [2].

3.3 MULTIVARIATE INCREASING CONVEX ORDERS

In this section, several extensions of the univariate increasing convex order are considered. These definitions depend on the kind of "convexity" that we assume on the functions involved in the definition. The first natural extension that we can consider is as follows.

Definition 3.3.1. Given two random vectors $\mathbf{X} = (X_1, \ldots, X_n)$ and $\mathbf{Y} = (Y_1, \ldots, Y_n)$, we say that $\mathbf{X}$ is smaller than $\mathbf{Y}$ in the *multivariate increasing convex order*, denoted by $\mathbf{X} \leq_{\mathrm{icx}} \mathbf{Y}$, if

$$E[\phi(\mathbf{X})] \leq E[\phi(\mathbf{Y})],$$

for all increasing convex function $\phi : \mathbb{R}^n \mapsto \mathbb{R}$, such that the previous expectations exist.

Assuming other kinds of convexity on the function ϕ, some additional definitions can be provided. Let us see some of them.

We shall say that a function $\phi : \mathbb{R}^n \mapsto \mathbb{R}$ is *componentwise convex* if is convex on each component when the other ones are held fixed. In particular, if ϕ is twice differentiable, then it is componentwise convex if $\partial^2 \phi / \partial x_i \partial x_i \geq 0$, for all $1 \leq i \leq n$. According to this notion, we have the following definition.

Definition 3.3.2. Given two random vectors $\mathbf{X} = (X_1, \ldots, X_n)$ and $\mathbf{Y} = (Y_1, \ldots, Y_n)$, we say that $\mathbf{X}$ is smaller than $\mathbf{Y}$ in the *multivariate increasing componentwise convex order*, denoted by $\mathbf{X} \leq_{\mathrm{iccx}} \mathbf{Y}$, if

$$E[\phi(\mathbf{X})] \leq E[\phi(\mathbf{Y})],$$

for all increasing componentwise convex function $\phi : \mathbb{R}^n \mapsto \mathbb{R}$, such that the previous expectations exist.

Let us recall next some other classes of functions defined on $\mathbb{R}^n$ related to the notion of convexity. The definitions are based on the notion of *difference operator*.

Let Δ_i^ϵ be the ith difference operator defined for a function $\phi : \mathbb{R}^n \mapsto \mathbb{R}$ as

$$\Delta_i^\epsilon \phi(\mathbf{x}) = \phi(\mathbf{x} + \epsilon \mathbf{1}_i) - \phi(\mathbf{x}),$$

where $\mathbf{1}_i = (0, \ldots, 0, \overset{i}{1}, 0, \ldots, 0)$. Then, a function ϕ is said to be *directionally convex* if $\Delta_i^\epsilon \Delta_j^\delta \phi(\mathbf{x}) \geq 0$, for all $1 \leq i \leq j \leq n$ and $\epsilon, \delta \geq 0$. Furthermore, if $\Delta_i^\epsilon \Delta_j^\delta \phi(\mathbf{x}) \geq 0$, for all $1 \leq i < j \leq n$ and $\epsilon, \delta \geq 0$, then ϕ is said to be *supermodular*.

Directionally convex functions are also known as ultramodular functions [113]. If ϕ is twice differentiable, then ϕ is directionally convex if $\partial^2 \phi / \partial x_i \partial x_j \geq 0$, for all $1 \leq i \leq j \leq n$. If $\partial^2 \phi / \partial x_i \partial x_j \geq 0$, for all

$1 \leq i < j \leq n$, then ϕ is supermodular. Besides, a function is directionally convex if it is supermodular and componentwise convex.

According to these properties, the following criteria are defined.

Definition 3.3.3. Given two random vectors $\mathbf{X} = (X_1, \ldots, X_n)$ and $\mathbf{Y} = (Y_1, \ldots, Y_n)$, we say that $\mathbf{X}$ is smaller than $\mathbf{Y}$ in the *multivariate increasing directionally convex [supermodular] order*, denoted by $\mathbf{X} \leq_{\text{idcx[ism]}} \mathbf{Y}$, if

$$E[\phi(\mathbf{X})] \leq E[\phi(\mathbf{Y})],$$

for all increasing directionally convex [supermodular] function $\phi : \mathbb{R}^n \mapsto \mathbb{R}$, such that the previous expectations exist.

The supermodular order is a well-known tool for comparing dependence structures of random vectors, whereas the directionally convex order compares not only the dependence structure but also the variability of the marginal distributions. Therefore, the ism order is not usually considered as an increasing convex order, and hence we shall restrict our attention to the icx, iccx and idcx orders.

From the relationships among the different types of convexities, the following relationships among the previous orders can be easily obtained:

$$\begin{array}{c} \mathbf{X} \leq_{\text{iccx}} \mathbf{Y} \;\; \Rightarrow \;\; \mathbf{X} \leq_{\text{idcx}} \mathbf{Y} \\ \Downarrow \\ \mathbf{X} \leq_{\text{icx}} \mathbf{Y} \end{array} \qquad (3.8)$$

From an applied point of view, we shall restrict our attention to the iccx and idcx orders. For example, the idcx order is interesting when we want to compare convolutions in terms of the univariate icx order. If we consider an increasing convex function $\psi : \mathbb{R} \mapsto \mathbb{R}$ and $\phi : \mathbb{R}^n \mapsto \mathbb{R}$ is an increasing directionally convex function, then the composition $\psi(\phi)$ is increasing directionally convex [113]. Taking into account previous comments and given that $\phi(x_1, \ldots, x_n) = \sum_{i=1}^n x_i$ is obviously increasing directionally convex, the following result is established.

Theorem 3.3.4. *Let* $\mathbf{X} = (X_1, \ldots, X_n)$ *and* $\mathbf{Y} = (Y_1, \ldots, Y_n)$ *be two random vectors. If* $\mathbf{X} \leq_{\text{idcx}} \mathbf{Y}$, *then*

$$\sum_{i=1}^n X_i \leq_{\text{icx}} \sum_{i=1}^n Y_i.$$

From (3.8), it is obvious that the previous result also holds if we assume $\mathbf{X} \leq_{\text{iccx}} \mathbf{Y}$.

Next, several results for these orders are provided. The proofs require several specific properties and notions that do not fit our objectives, and thus they are omitted. Proper references will be given.

The first result is a characterization of the iccx and idcx orders in terms of twice differentiable functions. This is useful because in some cases the proof can be reduced to the study of twice differentiable functions. For a detailed proof, we refer the reader to Ref. [114].

Theorem 3.3.5. *Let* $\mathbf{X} = (X_1, \ldots, X_n)$ *and* $\mathbf{Y} = (Y_1, \ldots, Y_n)$ *be two random vectors. Then,* $\mathbf{X} \leq_{\text{iccx[idcx]}} \mathbf{Y}$, *if, and only if,*

$$E[\phi(\mathbf{X})] \leq E[\phi(\mathbf{Y})],$$

for all twice differentiable increasing componentwise [directionally] convex function $\phi : \mathbb{R}^n \mapsto \mathbb{R}$, *such that the previous expectations exist.*

In order to provide examples where the idcx order holds, a result for random vectors with the same copula is given. We refer the reader to Ref. [115] for a proof, which is based on previous arguments and results by Müller and Scarsini [116].

Theorem 3.3.6. *Let* $\mathbf{X} = (X_1, \ldots, X_n)$ *and* $\mathbf{Y} = (Y_1, \ldots, Y_n)$ *be two random vectors with a common CI copula C. If* $X_i \leq_{\text{icx}} Y_i$, *for all* $i = 1, \ldots, n$, *then*

$$\mathbf{X} \leq_{\text{idcx}} \mathbf{Y}.$$

The previous result can be combined with Theorem 3.3.4 to provide the following result, which generalizes Theorem 2.3.13.

Theorem 3.3.7. *Let* $\mathbf{X} = (X_1, \ldots, X_n)$ *and* $\mathbf{Y} = (Y_1, \ldots, Y_n)$ *be two random vectors with a common CI copula C. If* $X_i \leq_{\text{icx}} Y_i$, *for all* $i = 1, \ldots, n$, *then*

$$\sum_{i=1}^{n} X_i \leq_{\text{icx}} \sum_{i=1}^{n} Y_i.$$

Finally, the following result for the comparison on the iccx order of random vectors with conditionally independent components is given. In this case, we consider a slightly different situation than the one considered for the multivariate stochastic order. Let $(X_1,\ldots,X_n,\Theta_1)$, $(Y_1,\ldots,Y_n,\Theta_2)$ be two random vectors, where $\Theta_1 = (\Theta_{11},\ldots,\Theta_{1n})$ and $\Theta_2 = (\Theta_{21},\ldots,\Theta_{2n})$ are two random vectors with a common support. Let us assume that the conditional distributions of X_i [and analogously for Y_i] given $\Theta_1 = \boldsymbol{\theta}$ [$\Theta_2 = \boldsymbol{\theta}$], for all $i = 1,\ldots,n$, just depends on $\Theta_{1i} = \theta_i$ [$\Theta_{2i} = \theta_i$]. In this case, we shall fix the notation $X_i(\theta_i) = X_i(\boldsymbol{\theta})$ [$Y_i(\theta_i) = Y_i(\boldsymbol{\theta})$]. Now, the following theorem can be stated.

Theorem 3.3.8. *According to the previous notation, assume that*

(i) $X_i(\theta) \leq_{\text{icx}} Y_i(\theta)$, for any θ in the corresponding support and for all $i = 1,\ldots,n$,

(ii) $E[\phi(X_i(\theta))]$ or $E[\phi(Y_i(\theta))]$ or both are increasing convex in θ, for all increasing convex function ϕ and for all $i = 1,\ldots,n$, and

(iii) $\Theta_1 \leq_{\text{iccx}} \Theta_2$.

Then,

$$(X_1,\ldots,X_n) \leq_{\text{iccx}} (Y_1,\ldots,Y_n).$$

Proof. Following Denuit and Müller [114], it is not difficult to show that whenever (ii) holds for $X_i(\theta)$, for all $i = 1,\ldots,n$, then $E[\phi(X_1(\theta_1),\ldots,X_n(\theta_n))]$ is increasing componentwise convex in $\boldsymbol{\theta}$ (the proof whenever (ii) holds for $Y_i(\theta)$, for all $i = 1,\ldots,n$, is similar). Now, (i) is equivalent to

$$(X_1(\theta_1),\ldots,X_n(\theta_n)) \leq_{\text{iccx}} (Y_1(\theta_1),\ldots,Y_n(\theta_n)),$$

and the proof follows under similar steps to those of Theorem 3.2.10.   □

Related results can be found in Refs. [112, 117].

3.4 MULTIVARIATE RESIDUAL LIFE ORDERS

In this section, multivariate generalizations of the univariate hazard rate and mean residual life orders defined in Section 2.4 are provided. These versions are based on the multivariate dynamic hazard rate and mean residual life functions given in Section 1.3. It is also possible to provide generalizations of the hazard rate order based on the ratio of the multivariate survival functions [118]. However, there are only a few results for these notions, and we restrict our attention to the previous ones. Let us proceed with the first definition.

Definition 3.4.1. Given two non-negative continuous random vectors $\mathbf{X} = (X_1, \ldots, X_n)$ and $\mathbf{Y} = (Y_1, \ldots, Y_n)$ with multivariate dynamic hazard rate functions $\eta.(\cdot|\cdot)$ and $\lambda.(\cdot|\cdot)$, respectively, we say that $\mathbf{X}$ is smaller than $\mathbf{Y}$ in the *multivariate dynamic hazard rate order*, denoted by $\mathbf{X} \leq_{\text{dyn-hr}} \mathbf{Y}$, if

$$\eta_k(t|h_t) \geq \lambda_k(t|h_t'), \quad \text{for all } t \geq 0,$$

for all

$$h_t = \{\mathbf{X}_{I \cup J} = \mathbf{x}_{I \cup J}, \mathbf{X}_{\overline{I \cup J}} > t\mathbf{e}\}$$

and

$$h_t' = \{\mathbf{Y}_I = \mathbf{y}_I, \mathbf{Y}_{\overline{I}} > t\mathbf{e}\},$$

whenever $I \cap J = \emptyset$, $\mathbf{0} \leq \mathbf{x}_I \leq \mathbf{y}_I \leq t\mathbf{e}$, $\mathbf{0} \leq \mathbf{x}_J \leq t\mathbf{e}$, and for all $k \in \overline{I \cup J}$. In this case, we say that h_t is *more severe* than h_t'.

Obviously, in the previous definition, it is implicitly assumed that it is possible to condition on the events h_t and h_t'.

The multivariate hazard rate order is not necessarily reflexive. In fact, if a random vector $\mathbf{X}$ satisfies $\mathbf{X} \leq_{\text{dyn-hr}} \mathbf{X}$, then it is said to have the HIF property (hazard increasing upon failure, see Ref. [119]) and it is considered a positive dependence property.

We wish to point out that the definition of the multivariate hazard rate order in the discrete case does not follow just considering dynamic discrete hazard rates in the previous definition. In this case, some additional considerations have to be taken into account [120, 121].

The multivariate dynamic hazard rate order is stronger than the multivariate stochastic order. Next, this result is stated without a proof. For a detailed proof, which requires the multivariate cumulative hazard order notion, the reader can look in Ref. [2].

Theorem 3.4.2. *Let* $\mathbf{X} = (X_1, \ldots, X_n)$ *and* $\mathbf{Y} = (Y_1, \ldots, Y_n)$ *be two non-negative continuous random vectors. If* $\mathbf{X} \leq_{\text{dyn-hr}} \mathbf{Y}$, *then*

$$\mathbf{X} \leq_{\text{st}} \mathbf{Y}.$$

In a similar way to Theorem 2.4.3, it is possible to characterize the dynamic hazard rate order in terms of comparisons of residual lives, as it is established next.

Theorem 3.4.3. *Let* $\mathbf{X} = (X_1, \ldots, X_n)$ *and* $\mathbf{Y} = (Y_1, \ldots, Y_n)$ *be two random vectors. Then,* $\mathbf{X} \leq_{\text{dyn-hr}} \mathbf{Y}$ *if, and only if,*

$$[(\mathbf{X} - t\mathbf{e})_+ | h_t] \leq_{\text{dyn-hr}} [(\mathbf{Y} - t\mathbf{e})_+ | h'_t],$$

for all h_t *and* h'_t, *where* h_t *is more severe than* h'_t.

Proof. The proof relies on the fact that given two histories h_t and h'_t, where h_t is more severe than h'_t, we see that conditioning the random vectors $[(\mathbf{X} - t\mathbf{e})_+ | h_t]$ and $[(\mathbf{Y} - t\mathbf{e})_+ | h'_t]$ on two histories h_s and h'_s, where h_s is more severe than h'_s, is equivalent to condition the random vectors $(\mathbf{X} - (t+s)\mathbf{e})_+$ and $(\mathbf{Y} - (t+s)\mathbf{e})_+$ on two histories h_{t+s} and h'_{t+s}, where h_{t+s} is more severe than h'_{t+s}. $\qquad \square$

There are not too many results on the preservation of the multivariate dynamic hazard rate order. For example, the dynamic hazard rate order is not preserved under marginalization, unless a dynamic conditional marginalization is considered [122].

As far as we know, increasing transformations of the whole random vectors do not preserve the dynamic hazard rate order. However, if we consider strictly increasing transformations of the components, it is possible to prove the following theorem.

Theorem 3.4.4. *Let* $\mathbf{X} = (X_1, \ldots, X_n)$ *and* $\mathbf{Y} = (Y_1, \ldots, Y_n)$ *be two non-negative continuous random vectors. If* $\mathbf{X} \leq_{\text{dyn-hr}} \mathbf{Y}$, *then*

$$(\phi(X_1), \ldots, \phi(X_n)) \leq_{\text{dyn-hr}} (\phi(Y_1), \ldots, \phi(Y_n)),$$

for all differentiable strictly increasing function $\phi : \mathbb{R} \mapsto \mathbb{R}$.

Proof. Let us consider the history

$$h_t = \left\{ \bigcap_{i \in I} \{\phi(X_i) = x_i\}, \bigcap_{j \in \bar{I}} \{\phi(X_j) > t\} \right\},$$

for the random vector $(\phi(X_1), \ldots, \phi(X_n))$. Let $\eta.(\cdot | \cdot)$ be the multivariate dynamic hazard rate associated to $\mathbf{X}$. Then, it is easy to see that the multivariate dynamic hazard rate of $\phi(X_j)$ given the history h_t, for all $j \in \bar{I}$, is given by

$$\left| \frac{1}{\phi'(\phi^{-1}(t))} \right| \eta_j \left(\phi^{-1}(t) \big| I_{\phi^{-1}(t)} \right),$$

where

$$l_{\phi^{-1}(t)} = \left\{ \bigcap_{i \in I} \{X_i = \phi^{-1}(x_i)\}, \bigcap_{j \in \bar{I}} \{X_j > \phi^{-1}(t)\} \right\}.$$

Analogously, a similar result for $(\phi(Y_1), \ldots, \phi(Y_n))$ holds. Now, the result follows from the assumption $X \leq_{\text{dyn-hr}} Y$. $\qquad\qquad\square$

Next, the multivariate dynamic hazard rate order for random vectors with conditionally independent components is studied. Let us consider two random vectors $(X_1, \ldots, X_n, \Theta_1)$ and $(Y_1, \ldots, Y_n, \Theta_2)$, where Θ_1 and Θ_2 are m-dimensional random vectors with common support S. Let $s_t^i(\theta)$ and $r_t^i(\theta)$ denote the hazard rate of $[X_i|\Theta_1 = \theta]$ and $[Y_i|\Theta_2 = \theta]$ at time $t > 0$, respectively, for all $i = 1, \ldots, n$. Let η and λ denote the multivariate dynamic hazard rate functions of $X = (X_1, \ldots, X_n)$ and $Y = (Y_1, \ldots, Y_n)$, respectively. Finally, let us suppose that X and Y are conditionally independent given $\Theta_1 = \theta$ and $\Theta_2 = \theta$, respectively, as in Theorem 3.2.10.

Theorem 3.4.5. *According to the previous notation, assume that*

(i) $[X_i|\Theta_1 = \theta]$ *(or* $[Y_i|\Theta_2 = \theta]$*) is increasing [decreasing] in the hazard rate order in* θ, *for all* $i = 1, \ldots, n$,

(ii) $[X_i|\Theta_1 = \theta] \leq_{\text{hr}} [Y_i|\Theta_2 = \theta]$, *for any* $\theta \in S$ *and for all* $i = 1, \ldots, n$, *and*

(iii) $[\Theta_1|h_t] \leq_{\text{st}} [\geq_{\text{st}}][\Theta_2|h_t']$, *for all histories* h_t *and* h_t', *for* $(X_1, \ldots, X_n)$ *and* $(Y_1, \ldots, Y_n)$, *respectively, where* h_t *is more severe than* h_t'.

Then,

$$(X_1, \ldots, X_n) \leq_{\text{dyn-hr}} (Y_1, \ldots, Y_n).$$

Proof. First, given two histories h_t and h_t' for X and Y, respectively, where h_t is more severe than h_t', let us see that the multivariate dynamic hazard rates for X and Y are given by

$$\eta_j(t|h_t) = \int_S s_t^j(\theta) h_1(\theta|h_t) \, d\theta, \qquad\qquad (3.9)$$

and

$$\lambda_j(t|h_t') = \int_S r_t^j(\theta) h_2(\theta|h_t') \, d\theta, \qquad\qquad (3.10)$$

respectively, where $h_i(\theta|D)$ denotes the conditional joint density function of Θ_i given a history D, for $i = 1, 2$.

Let $h_t = \{\mathbf{X}_I = \mathbf{x}_I, \mathbf{X}_{\bar{I}} > te\}$ be a history, where $I = \{i_1, \ldots, i_k\}$, $\bar{I} = \{j_1, \ldots, j_{n-k}\}$ and $j \in \bar{I}$. Then,

$$\eta_j(t|h_t) = \lim_{\Delta \to 0} \frac{1}{\Delta} P[t < X_j \leq t + \Delta|h_t]$$

$$= \lim_{\Delta \to 0} \frac{P[t < X_j \leq t + \Delta|\mathbf{X}_I = \mathbf{x}_I, \mathbf{X}_{\bar{I}} > t]}{\Delta}$$

$$= \lim_{\Delta \to 0} \frac{1}{\Delta} \frac{P[\mathbf{X}_{\bar{I}-j} > t, t < X_j \leq t + \Delta|\mathbf{X}_I = \mathbf{x}_I]}{P[\mathbf{X}_{\bar{I}} > te|\mathbf{X}_I = \mathbf{x}_I]}.$$

Let us fix some notation. Let $f_i(\cdot|\theta)$ and $F_i(\cdot|\theta)$ denote the density and distribution functions of $[X_i|\Theta_1 = \theta]$, respectively. Let f denotes the joint density of $(\mathbf{X}_I, \mathbf{X}_{\bar{I}})$, f_I the joint density of $\mathbf{X}_I$ and h_1 the joint density of Θ_1. According to this notation, we see that

$$P[\mathbf{X}_{\bar{I}-j} > te, t < X_j \leq t + \Delta|\mathbf{X}_I = \mathbf{x}_I]$$

$$= \frac{\int_t^{+\infty} \cdots \int_t^{t+\Delta} \cdots \int_t^{+\infty} f(\mathbf{x}_I, \mathbf{x}_{\bar{I}}) \, dx_{j_1} \ldots, dx_{j_{n-k}}}{f_I(x_{i_1}, \ldots, x_{i_k})}$$

$$= \frac{\int_S \prod_{i\in I} f_i(x_i|\theta) \prod_{l\in\bar{I}-j} \overline{F}_l(t|\theta)[F_j(t+\Delta|\theta) - F_j(t|\theta)]h_1(\theta) \, d\theta}{\int_S \prod_{i\in I} f_i(x_i|\theta)h_1(\theta) \, d\theta},$$

and, analogously,

$$P[\mathbf{X}_{\bar{I}} > te|\mathbf{X}_I = \mathbf{x}_I] = \frac{\int_S \prod_{i\in I} f_i(x_i|\theta) \prod_{l\in\bar{I}} \overline{F}_l(t|\theta)h_1(\theta) \, d\theta}{\int_S \prod_{i\in I} f_i(x_i|\theta)h_1(\theta) \, d\theta}.$$

Replacing the last two equalities in $\eta_j(t|h_t)$, changing the order of integration, and using the Lebesgue's dominated convergence theorem, we see that

$$\eta_j(t|h_t) = \lim_{\Delta \to 0}$$

$$\frac{1}{\Delta} \frac{\int_S \prod_{i\in I} f_i(x_i|\theta) \prod_{l\in\bar{I}-j} \overline{F}_l(t|\theta)[F_j(t+\Delta|\theta) - F_j(t|\theta)]h_1(\theta) \, d\theta}{\int_S \prod_{i\in I} f_i(x_i|\theta) \prod_{l\in\bar{I}} \overline{F}_l(t|\theta)h_1(\theta) \, d\theta}$$

$$= \lim_{\Delta \to 0}$$

$$\frac{1}{\Delta} \int_S \frac{F_j(t+\Delta|\boldsymbol{\theta}) - F_j(t|\boldsymbol{\theta})}{\overline{F}_j(t|\boldsymbol{\theta})} \frac{\prod_{i\in I} f_i(x_i|\boldsymbol{\theta}) \prod_{l\in \overline{I}} \overline{F}_l(t|\boldsymbol{\theta}) h_1(\boldsymbol{\theta}) \, \mathrm{d}\boldsymbol{\theta}}{\int_S \prod_{i\in I} f_i(x_i|\boldsymbol{\theta}) \prod_{l\in \overline{I}} \overline{F}_l(t|\boldsymbol{\theta}) h_1(\boldsymbol{\theta}) \, \mathrm{d}\boldsymbol{\theta}} \, \mathrm{d}\boldsymbol{\theta}$$

$$= \int_S \frac{1}{\overline{F}_j(t|\boldsymbol{\theta})} \left(\lim_{\Delta \to 0} \frac{F_j(t+\Delta|\boldsymbol{\theta}) - F_j(t|\boldsymbol{\theta})}{\Delta} \right) h_1(\boldsymbol{\theta}|h_t) \, \mathrm{d}\boldsymbol{\theta}$$

$$= \int_S s_t^j(\boldsymbol{\theta}) h_1(\boldsymbol{\theta}|h_t) \, \mathrm{d}\boldsymbol{\theta}.$$

Let us prove the result. Observe that condition (i) is equivalent to the condition $s_t^i(\boldsymbol{\theta})$ (or $r_t^i(\boldsymbol{\theta})$) is decreasing [increasing] in $\boldsymbol{\theta}$, for all $t > 0$. We will consider the case in which $s_t^i(\boldsymbol{\theta})$ is decreasing [increasing] in $\boldsymbol{\theta}$, for all $t > 0$, the other case follows under similar arguments.

Let h_t and h_t' be two histories as in Definition 3.4.1 and $j \in \overline{I \cup J}$. Recalling (3.9) and (3.10), we get

$$\eta_j(t|h_t) = \int_S s_t^j(\boldsymbol{\theta}) h_1(\boldsymbol{\theta}|h_t) \, \mathrm{d}\boldsymbol{\theta} = E[s_t^j(\boldsymbol{\Theta}_1|h_t)]$$

$$\geq E[s_t^j(\boldsymbol{\Theta}_2|h_t')] \geq E[r_t^j(\boldsymbol{\Theta}_2|h_t')] = \lambda_j(t|h_t'),$$

where the first inequality follows from (i) and (iii), and the second inequality follows from (ii). Therefore, $(X_1, \ldots, X_n) \leq_{\text{dyn-hr}} (Y_1, \ldots, Y_n)$. □

Considering the multivariate dynamic mean residual lives (instead of the multivariate dynamic hazard rates) leads to a multivariate version of the mean residual life order.

Definition 3.4.6. Given two non-negative continuous random vectors $\mathbf{X} = (X_1, \ldots, X_n)$ and $\mathbf{Y} = (Y_1, \ldots, Y_n)$ with multivariate dynamic mean residual life functions $m.(\cdot|\cdot)$ and $l.(\cdot|\cdot)$, respectively, we say that $\mathbf{X}$ is smaller than $\mathbf{Y}$ in the *multivariate mean residual life order*, denoted by $\mathbf{X} \leq_{\text{mrl}} \mathbf{Y}$, if

$$m_k(t|h_t) \leq l_k(t|h_t'), \quad \text{for all } t \geq 0,$$

for all

$$h_t = \{\mathbf{X}_{I\cup J} = \mathbf{x}_{I\cup J}, \mathbf{X}_{\overline{I\cup J}} > t\mathbf{e}\},$$

and

$$h_t' = \{\mathbf{Y}_I = \mathbf{y}_I, \mathbf{Y}_{\overline{I}} > t\mathbf{e}\},$$

whenever $I \cap J = \emptyset$, $\mathbf{0} \leq \mathbf{x}_I \leq \mathbf{y}_I \leq t\mathbf{e}$, $\mathbf{0} \leq \mathbf{x}_J \leq t\mathbf{e}$, and for all $k \in \overline{I \cup J}$.

Again, the multivariate mean residual life order is not necessarily reflexive. If a random vector X satisfies $X \leq_{\text{mrl}} X$, then it is said to have the MRL-DF property (mean residual life decreasing upon failure, see Ref. [119]) and it is also considered a positive dependence property.

As we see next, the multivariate dynamic mean residual life order is weaker than the multivariate dynamic hazard rate order.

Theorem 3.4.7. *Let* $\mathbf{X} = (X_1, \ldots, X_n)$ *and* $\mathbf{Y} = (Y_1, \ldots, Y_n)$ *be two non-negative continuous random vectors. If* $\mathbf{X} \leq_{\text{dyn-hr}} \mathbf{Y}$, *then*

$$\mathbf{X} \leq_{\text{mrl}} \mathbf{Y}.$$

Proof. Let us consider two histories h_t and h'_t for $\mathbf{X}$ and $\mathbf{Y}$, respectively, as in Definition 3.4.6. Then, by Theorems 3.4.2 and 3.4.3, $[(\mathbf{X} - t\mathbf{e})_+ | h_t] \leq_{\text{st}}$ $[(\mathbf{Y} - t\mathbf{e})_+ | h'_t]$ holds. Now, from the preservation of the multivariate stochastic order under marginalization, we see that

$$[X_k - t | h_t] \leq_{\text{st}} [Y_k - t | h'_t],$$

for all $k \in \overline{I} \cup \overline{J}$. From (2.2), $E[X_k - t | h_t] \leq E[Y_k - t | h'_t]$ holds and, therefore, $\mathbf{X} \leq_{\text{mrl}} \mathbf{Y}$. $\qquad\qquad\square$

Now, a result for the comparison of random vectors with conditionally independent components is provided. According to the notation introduced for Theorem 3.4.5 and denoting by $m^i_t(\boldsymbol{\theta})$ and $l^i_t(\boldsymbol{\theta})$ the mean residual life of $[X_i | \boldsymbol{\Theta}_1 = \boldsymbol{\theta}]$ and $[Y_i | \boldsymbol{\Theta}_2 = \boldsymbol{\theta}]$, respectively, the following result is stated.

Theorem 3.4.8. *Assume that*

(i) $[X_i | \boldsymbol{\Theta}_1 = \boldsymbol{\theta}]$ *(or* $[Y_i | \boldsymbol{\Theta}_2 = \boldsymbol{\theta}]$*) is increasing [decreasing] in the mean residual life order in* $\boldsymbol{\theta}$, *for all* $i = 1, \ldots, n$,

(ii) $[X_i | \boldsymbol{\Theta}_1 = \boldsymbol{\theta}] \leq_{\text{mrl}} [Y_i | \boldsymbol{\Theta}_2 = \boldsymbol{\theta}]$, *for any* $\boldsymbol{\theta} \in S$ *and for all* $i = 1, \ldots, n$, *and*

(iii) $[\boldsymbol{\Theta}_1 | h_t] \leq_{\text{st}} [\geq_{\text{st}}][\boldsymbol{\Theta}_2 | h'_t]$, *for all histories* h_t *and* h'_t, *for* $(X_1, \ldots, X_n)$ *and* $(Y_1, \ldots, Y_n)$, *respectively, where* h_t *is more severe than* h'_t.

Then,

$$(X_1, \ldots, X_n) \leq_{\text{mrl}} (Y_1, \ldots, Y_n).$$

Proof. First, an alternative expression for the mean residual life of $\mathbf{X}$ is derived. Let us consider the history $h_t = \{X_I = x_I, X_{\overline{I}} > t\mathbf{e}\}$, where $I = \{i_1, \ldots, i_k\}, \overline{I} = \{j_1, \ldots, j_{n-k}\}$ and let $j \in \overline{I}$. Then,

$$m_j(t|h_t) = E[X_j - t|h_t] = E[E[X_j - t|\boldsymbol{\Theta}_1]|h_t] = E[m_t^i(\boldsymbol{\Theta}_1)|h_t], \quad (3.11)$$

where the last inequality follows from the conditional independence of $X_1, \ldots, X_n$ given $\boldsymbol{\Theta}_1$. A similar result for the multivariate dynamic mean residual life of $\mathbf{Y}$ holds. In particular,

$$l_j(t|h_t') = E[Y_j - t|h_t'] = E[l_t^i(\boldsymbol{\Theta}_2)|h_t']. \quad (3.12)$$

The proof now follows under similar steps to those of Theorem 3.4.5. First, observe that condition (i) is equivalent to $m_t^i(\boldsymbol{\theta})$ (or $l_t^i(\boldsymbol{\theta})$) being increasing [decreasing] in $\boldsymbol{\theta}$, for all $t > 0$. Let us suppose that $m_t^i(\boldsymbol{\theta})$ is increasing [decreasing] in $\boldsymbol{\theta}$, for all $t > 0$ (the other case is similar).

Let us consider two histories h_t and h_t' for $(X_1, \ldots, X_n)$ and $(Y_1, \ldots, Y_n)$, respectively, such that h_t is more severe than h_t'. Recalling (3.11) and (3.12), we see that

$$m_i(t|h_t) = \int_S m_t^i(\boldsymbol{\theta}) h_1(\boldsymbol{\theta}|h_t) \, d\boldsymbol{\theta} = E[m_t^i(\boldsymbol{\Theta}_1|h_t)]$$

$$\leq E[m_t^i(\boldsymbol{\Theta}_2|h_t')]$$

$$\leq E[l_t^i(\boldsymbol{\Theta}_2|h_t')] = l_i(t|h_t'),$$

where the first inequality follows from (i) and (iii), and the second inequality follows from (ii). Therefore, $(X_1, \ldots, X_n) \leq_{\text{mrl}} (Y_1, \ldots, Y_n)$. $\quad\square$

3.5 THE MULTIVARIATE LIKELIHOOD RATIO ORDER

In a similar way to the univariate case, it is possible to check the multivariate stochastic and hazard rate orders in terms of a property of the joint density functions. This property leads to the definition of the multivariate likelihood ratio order, which is a generalization of the univariate likelihood ratio order. In this section, the results are given in the continuous case but they can be restated for the general case. The main intention of the multivariate likelihood ratio order is to provide a sufficient condition for the multivariate hazard rate order.

Definition 3.5.1. Given two continuous random vectors $\mathbf{X} = (X_1, \ldots, X_n)$ and $\mathbf{Y} = (Y_1, \ldots, Y_n)$ with joint density functions f and g, respectively, we say that $\mathbf{X}$ is smaller than $\mathbf{Y}$ in the *multivariate likelihood ratio order*, denoted by $\mathbf{X} \leq_{\text{lr}} \mathbf{Y}$, if

$$f(\mathbf{x})g(\mathbf{y}) \le f(\mathbf{x} \wedge \mathbf{y})g(\mathbf{x} \vee \mathbf{y}), \quad \text{for all } \mathbf{x}, \mathbf{y} \in \mathbb{R}^n.$$

Clearly, this is a generalization of the likelihood ratio order in the univariate case. However, the multivariate likelihood ratio order is not necessarily reflexive. When a random vector $\mathbf{X}$ satisfies $\mathbf{X} \le_{lr} \mathbf{X}$, we have the MTP$_2$ property introduced in Section 1.3.

The following result provides a set of sufficient conditions for the multivariate likelihood ratio order.

Theorem 3.5.2. *Let* $\mathbf{X} = (X_1, \ldots, X_n)$ *and* $\mathbf{Y} = (Y_1, \ldots, Y_n)$ *be two continuous random vectors with joint density functions f and g, respectively. If* $\mathbf{X}$ *or* $\mathbf{Y}$ *or both are MTP$_2$, and*

$$f(\mathbf{y})g(\mathbf{x}) \le f(\mathbf{x})g(\mathbf{y}), \quad \text{for all } \mathbf{x}, \mathbf{y} \in \mathbb{R}^n, \text{ such that } \mathbf{x} \le \mathbf{y}, \qquad (3.13)$$

then

$$\mathbf{X} \le_{lr} \mathbf{Y}.$$

Proof. Let us assume that $\mathbf{X}$ is MTP$_2$ (in the other case the proof is similar), then we have the following inequalities:

$$f(\mathbf{x})g(\mathbf{y}) \le \frac{f(\mathbf{x} \wedge \mathbf{y})f(\mathbf{x} \vee \mathbf{y})}{f(\mathbf{y})}g(\mathbf{y}) \le f(\mathbf{x} \wedge \mathbf{y})g(\mathbf{x} \vee \mathbf{y}),$$

for all $\mathbf{x}, \mathbf{y} \in \mathbb{R}^n$, such that $\mathbf{x} \le \mathbf{y}$, where the first inequality follows from the MTP$_2$ property and the second one from (3.13). Therefore, $\mathbf{X} \le_{lr} \mathbf{Y}$. □

The multivariate likelihood ratio order is preserved under conditioning on sublattices, as we see next. Recall that a subset $A \subseteq \mathbb{R}^n$ is called a *sublattice* if $\mathbf{x}, \mathbf{y} \in A$ implies $\mathbf{x} \wedge \mathbf{y} \in A$ and $\mathbf{x} \vee \mathbf{y} \in A$. This result will be used to show the relationship among the multivariate likelihood ratio order and the multivariate dynamic hazard rate order. The proof is obvious from the definition.

Theorem 3.5.3. *Let* $\mathbf{X} = (X_1, \ldots, X_n)$ *and* $\mathbf{Y} = (Y_1, \ldots, Y_n)$ *be two continuous random vectors. If* $\mathbf{X} \le_{lr} \mathbf{Y}$, *then*

$$[\mathbf{X}|\mathbf{X} \subseteq A] \le_{lr} [\mathbf{Y}|\mathbf{Y} \subseteq A], \text{ for all sublattice } A \subseteq \mathbb{R}^n.$$

In particular, from the previous theorem, the multivariate likelihood ratio order is preserved under marginalization. This result is useful because, in some cases, it is easier to provide a result in the multivariate case rather that in the univariate case.

Theorem 3.5.4. *Let* $\mathbf{X} = (X_1, \ldots, X_n)$ *and* $\mathbf{Y} = (Y_1, \ldots, Y_n)$ *be two continuous random vectors. If* $\mathbf{X} \leq_{\text{lr}} \mathbf{Y}$, *then*

$$\mathbf{X}_I \leq_{\text{lr}} \mathbf{Y}_I, \quad \textit{for all } I \subseteq \{1, \ldots, n\}.$$

Next, it is showed that the multivariate likelihood ratio order is stronger than the multivariate hazard rate order.

Theorem 3.5.5. *Let* $\mathbf{X} = (X_1, \ldots, X_n)$ *and* $\mathbf{Y} = (Y_1, \ldots, Y_n)$ *be two continuous random vectors. If* $\mathbf{X} \leq_{\text{lr}} \mathbf{Y}$, *then*

$$\mathbf{X} \leq_{\text{hr}} \mathbf{Y}.$$

Proof. Let us check the conditions for the definition of the multivariate dynamic hazard rate order. In particular, denoting by η and λ the multivariate dynamic hazard rates of $\mathbf{X}$ and $\mathbf{Y}$, respectively, let us see if

$$\eta_k(t|h_t) \geq \lambda_k(t|h_t'), \quad \text{for all } t \geq 0,$$

where

$$h_t = \{\mathbf{X}_{I \cup J} = \mathbf{x}_{I \cup J}, \mathbf{X}_{\overline{I \cup J}} > t\mathbf{e}\},$$

and

$$h_t' = \{\mathbf{Y}_I = \mathbf{y}_I, \mathbf{Y}_{\overline{I}} > t\mathbf{e}\},$$

whenever $I \cap J = \emptyset$, $\mathbf{0} \leq \mathbf{x}_I \leq \mathbf{y}_I \leq t\mathbf{e}$, $\mathbf{0} \leq \mathbf{x}_J \leq t\mathbf{e}$, and for all $k \in \overline{I \cup J}$.

The result will follow by proving that, as we shall see later,

$$\left[\mathbf{X}_{\overline{I \cup J}} \middle| h_t \right] \leq_{\text{lr}} \left[\mathbf{Y}_{\overline{I \cup J}} \middle| h_t' \right]. \tag{3.14}$$

Condition (3.14) will follow if

$$\left[\mathbf{X}_{\overline{I \cup J}} \middle| \mathbf{X}_{I \cup J} = \mathbf{x}_{I \cup J} \right] \leq_{\text{lr}} \left[\mathbf{Y}_{\overline{I \cup J}} \middle| \mathbf{Y}_I = \mathbf{y}_I, \mathbf{Y}_J > t\mathbf{e} \right], \tag{3.15}$$

holds.

Let us see that (3.15) follows if $\mathbf{X} \leq_{\text{lr}} \mathbf{Y}$ holds. Denoting by f and g the joint density functions of $(\mathbf{X}_I, \mathbf{X}_J, \mathbf{X}_{I \cup J})$ and $(\mathbf{Y}_I, \mathbf{Y}_J, \mathbf{Y}_{I \cup J})$, respectively, and by $f_{I,J}$ and $g_{I,J}$ the joint densities of $\mathbf{X}_{I,J}$ and $\mathbf{Y}_{I,J}$, respectively, we see that the joint density function of $\left[\mathbf{X}_{\overline{I \cup J}} \middle| \mathbf{X}_{I \cup J} = \mathbf{x}_{I \cup J} \right]$ is given by

$$f_{\overline{I \cup J}}(\mathbf{x}_{\overline{I \cup J}}) = \frac{f(\mathbf{x}_I, \mathbf{x}_J, \mathbf{x}_{\overline{I \cup J}})}{f_{I,J}(\mathbf{x}_I, \mathbf{x}_J)},$$

and the joint density function of $\left[\mathbf{Y}_{\overline{I \cup J}} \middle| \mathbf{Y}_I = \mathbf{y}_I, \mathbf{Y}_J > t\mathbf{e} \right]$ is given by

$$g_{\overline{I\cup J}}(\mathbf{x}_{\overline{I\cup J}}) = \frac{\int_{\mathbf{y}_J > t\mathbf{e}} g(\mathbf{y}_I, \mathbf{y}_J, \mathbf{x}_{\overline{I\cup J}}) d\mathbf{y}_J}{\int_{\mathbf{y}_J > t\mathbf{e}} g_{I,J}(\mathbf{y}_I, \mathbf{y}_J) d\mathbf{y}_J}.$$

Given $\mathbf{y}_J > t\mathbf{e}$, we see that $\mathbf{x}_J < t\mathbf{e} < \mathbf{y}_J$ and, analogously, $\mathbf{x}_I \le \mathbf{y}_I$. Since $\mathbf{X} \le_{\text{lr}} \mathbf{Y}$, given $\mathbf{y}_{\overline{I\cup J}}$, we see that

$$f(\mathbf{x}_I, \mathbf{x}_J, \mathbf{x}_{\overline{I\cup J}}) g(\mathbf{y}_I, \mathbf{y}_J, \mathbf{y}_{\overline{I\cup J}})$$

$$\le f(\mathbf{x}_I, \mathbf{x}_J, \mathbf{x}_{\overline{I\cup J}} \wedge \mathbf{y}_{\overline{I\cup J}}) g(\mathbf{y}_I, \mathbf{y}_J, \mathbf{x}_{\overline{I\cup J}} \vee \mathbf{y}_{\overline{I\cup J}}),$$

which, upon integration, yields

$$f_{\overline{I\cup J}}(\mathbf{x}_{\overline{I\cup J}}) g_{\overline{I\cup J}}(\mathbf{y}_{\overline{I\cup J}}) \le f_{\overline{I\cup J}}(\mathbf{x}_{\overline{I\cup J}} \wedge \mathbf{y}_{\overline{I\cup J}}) g_{\overline{I\cup J}}(\mathbf{x}_{\overline{I\cup J}} \vee \mathbf{y}_{\overline{I\cup J}}).$$

Therefore, from previous inequality, we see that (3.15) holds. Now, from Theorem 3.5.3, we see that (3.14) also holds. In particular, $[X_k | h_t] \le_{\text{lr}} [Y_k | h_t']$ holds, for all $k \in \overline{I\cup J}$. Now, since (2.25), we see that the hazard rates of $[X_k | h_t]$ and $[Y_k | h_t']$ are ordered and, consequently, we see that

$$\eta_k(t|h_t) \ge \lambda_k(t|h_t').$$

$\square$

Finally, a result for the multivariate likelihood ratio order among random vectors with conditionally independent components is provided. In particular, we consider the same background as that in Theorem 3.3.8.

Theorem 3.5.6. *Assume that*

(i) $X_i(\theta) =_{\text{st}} Y_i(\theta)$, for all θ and for all $i = 1, \ldots, n$,
(ii) $X_i(\theta) \le_{\text{lr}} Y_j(\theta')$, for all $\theta \le \theta'$ and for all $1 \le i \le j \le n$, and
(iii) $\Theta_1 \le_{\text{lr}} \Theta_2$.

Then,

$$(X_1, \ldots, X_n) \le_{\text{lr}} (Y_1, \ldots, Y_n).$$

Proof. Let $f_i(x|\theta)$ be the density function of $X_i(\theta)$. From condition (ii), we see that

$$\prod_{t=1}^n f_i(x_i|\theta_i) \text{ is TP}_2 \text{ in } (x_1, \ldots, x_n, \theta_1, \ldots, \theta_n).$$

Furthermore, condition (iii) is equivalent to the fact that $h_i(\theta_1, \ldots, \theta_n)$ is TP$_2$ in $(\theta_1, \ldots, \theta_n, i) \in S \times \{1, 2\}$. Due to these facts and from Theorem 1.2.3, we see that $\int f(x_1,\ldots,x_n|\theta_1,\ldots,\theta_n)dH_i(\boldsymbol{\theta})$ is TP_2 in $(x_1,\ldots,x_n,i)$, and we get the result. $\qquad\qquad\qquad\qquad\qquad\qquad\qquad\qquad\qquad\qquad\qquad$ $\square$

3.6 THE MULTIVARIATE DISPERSIVE ORDER

The definition of the univariate dispersive order proposed by Lewis and Thompson [64] (see Section 2.6) is based on the property of a random variable having quantiles more widely spread. In the literature, there are different extensions of this order (see, for instance, Refs. [32, 123]). In particular, we recall here the generalization proposed by Fernández-Ponce and Suárez-Llorens [32] by means of the so-called standard construction (see Section 1.3).

Definition 3.6.1. Given two continuous random vectors $\mathbf{X} = (X_1,\ldots, X_n)$ and $\mathbf{Y} = (Y_1,\ldots,Y_n)$ with standard constructions $\mathbf{Q_X}$ and $\mathbf{Q_Y}$, respectively, we say that $\mathbf{X}$ is smaller than $\mathbf{Y}$ in *multivariate dispersive order*, denoted as $\mathbf{X} \leq_{\mathrm{disp}} \mathbf{Y}$, if

$$\|\mathbf{Q_Y(p)} - \mathbf{Q_Y(q)}\|_2 \leq \|\mathbf{Q_X(p)} - \mathbf{Q_X(q)}\|_2,$$

for all $\mathbf{p},\mathbf{q} \in (0,1)^n$, where $\|\cdot\|_2$ is the Euclidean distance in $\mathbb{R}^n$.

As will be shown, the multivariate dispersive order introduced by Fernández-Ponce and Suárez-Llorens [32] has a good characterization in terms of a multivariate dispersive transformation between the random vectors.

First, we recall the definition of a multivariate dispersive function. Let $\mathbf{f} : \mathbb{R}^n \mapsto \mathbb{R}^n$, it is said that $\mathbf{f}$ is a dispersive function on its support if

$$\|\mathbf{x} - \mathbf{y}\|_2 \leq \|\mathbf{f(x)} - \mathbf{f(y)}\|_2, \quad \text{for all } \mathbf{x}, \mathbf{y} \in \mathbb{R}^n.$$

If $\mathbf{f}$ is differentiable, then $\mathbf{f}$ is a dispersive function if, and only if,

$$I_n \leq_L J_{\mathbf{f}}(\mathbf{x})^t J_{\mathbf{f}}(\mathbf{x}), \quad \text{for all } \mathbf{x} \in \mathbb{R}^n,$$

where

$$J_{\mathbf{f}}(\mathbf{x}) = \begin{pmatrix} \frac{\partial \mathbf{f}_1}{\partial x_1}(\mathbf{x}) & \frac{\partial \mathbf{f}_1}{\partial x_2}(\mathbf{x}) & \cdots & \frac{\partial \mathbf{f}_1}{\partial x_n}(\mathbf{x}) \\ \frac{\partial \mathbf{f}_2}{\partial x_1}(\mathbf{x}) & \frac{\partial \mathbf{f}_2}{\partial x_2}(\mathbf{x}) & \cdots & \frac{\partial \mathbf{f}_2}{\partial x_n}(\mathbf{x}) \\ \vdots & \vdots & \ddots & \vdots \\ \frac{\partial \mathbf{f}_n}{\partial x_1}(\mathbf{x}) & \frac{\partial \mathbf{f}_n}{\partial x_2}(\mathbf{x}) & \cdots & \frac{\partial \mathbf{f}_n}{\partial x_n}(\mathbf{x}) \end{pmatrix}$$

is the Jacobian matrix of $\mathbf{f}$, $J_{\mathbf{f}}(\mathbf{x})^t$ denotes its transposition, I_n is the identity matrix of order n and $\leq_L$ is the Loewner ordering of matrices, that means, $A \leq_L B$ if, and only if, $B - A$ is a non-negative definite matrix.

Recall also that, given two random vectors $\mathbf{X}$ and $\mathbf{Y}$, $\Phi = \mathbf{Q_Y}(\mathbf{D_X})$ satisfies that $\mathbf{Y} =_{st} \Phi(\mathbf{X})$ (see Section 1.3). The following theorem is a clear multivariate generalization of Theorem 2.6.3.

Theorem 3.6.2. *Let* $\mathbf{X} = (X_1, \ldots, X_n)$ *and* $\mathbf{Y} = (Y_1, \ldots, Y_n)$ *be two continuous random vectors, which supports are open intervals in* $\mathbb{R}^n$. *Then,* $\mathbf{X} \leq_{\text{disp}} \mathbf{Y}$ *if, and only if,* Φ *is a multivariate dispersive function.*

Proof. First, if $\mathbf{X} \leq_{\text{disp}} \mathbf{Y}$, then

$$||\Phi(\mathbf{x}) - \Phi(\mathbf{y})||_2 = ||\mathbf{Q_Y}(\mathbf{D_X}(\mathbf{x})) - \mathbf{Q_Y}(\mathbf{D_X}(\mathbf{y}))||_2$$

$$\geq ||\mathbf{Q_X}(\mathbf{D_X}(\mathbf{x})) - \mathbf{Q_X}(\mathbf{D_X}(\mathbf{y}))||_2 = ||\mathbf{x} - \mathbf{y}||_2,$$

for all $\mathbf{x}, \mathbf{y}$ in the support of $\mathbf{X}$. Hence, Φ is a dispersive function.

On the other hand, let us suppose that Φ is a multivariate dispersive function. Then,

$$||\mathbf{Q_X}(\mathbf{p}) - \mathbf{Q_X}(\mathbf{q})||_2$$

$$\leq ||\Phi(\mathbf{Q_X}(\mathbf{p})) - \Phi(\mathbf{Q_X}(\mathbf{q}))||_2 = ||\mathbf{Q_Y}(\mathbf{p}) - \mathbf{Q_Y}(\mathbf{q})||_2,$$

for all $\mathbf{p}, \mathbf{q} \in (0, 1)^n$, which means that $\mathbf{X} \leq_{\text{disp}} \mathbf{Y}$.   □

Therefore, roughly speaking, $\mathbf{X} \leq_{\text{disp}} \mathbf{Y}$ means that there exists a multivariate dispersive function Φ, from the support of $\mathbf{X}$ onto the support of $\mathbf{Y}$, such that

$$\mathbf{Y} =_{st} \Phi(\mathbf{X}),$$

for all $i = 1, \ldots, n$. Moreover, if it is assumed that Φ is differentiable, then its Jacobian matrix of Φ satisfies

$$I_n \leq_L J_\Phi(\mathbf{x})^t J_\Phi(\mathbf{x}), \quad \text{for all } \mathbf{x} \in \mathbb{R}^n,$$

and

$$\frac{\partial \Phi_i(x_1, \ldots, x_i)}{\partial x_i} \geq 0.$$

Moreover, if this is the case, then

$$\Phi_i(x_1, \ldots, x_i) = \mathbf{Q}_{Y,i}(\mathbf{D}_{X,i}(x_1, \ldots, x_i)).$$

Next, an example dealing with elliptically contoured distributions is shown (see Section 1.3).

Example 3.6.3. Let $\mathbf{X} \sim E_n(\mu_1, \mathbf{\Sigma}_1, g)$ and $\mathbf{Y} \sim E_n(\mu_2, \mathbf{\Sigma}_2, g)$ with a common generator g. According to Theorem 14.5.11 in Ref. [124], there exist two lower triangular matrices A and B, such that $AA^t = \mathbf{\Sigma}_2$ and $B^tB = \mathbf{\Sigma}_1^{-1}$ can be found. Furthermore, it holds that

$$A^t = \mathbf{D}_A^{1/2}\mathbf{U} \text{ and } B = \mathbf{D}_B^{-1/2}(\mathbf{V}^{-1})^t,$$

with $\mathbf{U}$ and $\mathbf{V}$ being the unique unit upper triangular matrices and $\mathbf{D}_A = \{d_{Ai}\}$ and $\mathbf{D}_B = \{d_{Bi}\}$ being the unique diagonal matrices with positive elements, such that

$$\mathbf{\Sigma}_1 = \mathbf{V}^t\mathbf{D}_B\mathbf{V} \text{ and } \mathbf{\Sigma}_2 = \mathbf{U}^t\mathbf{D}_A\mathbf{U},$$

where $\mathbf{D}_A^{1/2} = \{\sqrt{d_{Ai}}\}$ and similarly for $\mathbf{D}_B^{1/2}$. The $\mathbf{U}$ and $\mathbf{V}$ matrices are computed using the Cholesky decomposition, see Ref. [124].

From the properties of $\mathbf{\Phi}$ and the well-known fact that elliptically contoured distributions are preserved by affine transformations, we see that

$$\mathbf{\Phi}(\mathbf{x}) = \mathbf{Q}_\mathbf{Y}(\mathbf{D}_\mathbf{X}(\mathbf{x})) = AB(\mathbf{x} - \mu_1) + \mu_2.$$

Consequently, the Jacobian matrix of $\mathbf{\Phi}$ satisfies $J_\mathbf{\Phi} = AB$. Due to the fact that the Jacobian matrix is constant, it follows directly that $\mathbf{X} \leq_{\text{disp}} \mathbf{Y}$ if, and only if, $I_n \leq (AB)^t(AB)$.

The dependence structure of a random vector is an important aspect to take into account. As we see next, when the random vectors share the same copula, the univariate dispersive order among the marginal distributions implies the multivariate dispersive order among the random vectors.

Theorem 3.6.4. *Let* $\mathbf{X} = (X_1, \ldots, X_n)$ *and* $\mathbf{Y} = (Y_1, \ldots, Y_n)$ *be two continuous random vectors with a differentiable common copula C and differentiable marginal distribution functions. Then,* $\mathbf{X} \leq_{\text{disp}} \mathbf{Y}$ *if, and only if,* $X_i \leq_{\text{disp}} Y_i$, *for all* $i = 1, \ldots, n$.

Proof. Let F_i and G_i denote the distribution function of X_i and Y_i, respectively, for all $i = 1, \ldots, n$, and let $(U_1, \ldots, U_n)$ be the random vector with joint distribution function C. From the properties of the copula, we see that

$$(U_1, \ldots, U_n) =_{\text{st}} (F_1(X_1), \ldots, F_1(X_1)),$$
$$(Y_1, \ldots, Y_n) =_{\text{st}} (G_1^{-1}(U_1), \ldots, G_n^{-1}(U_1)).$$

Therefore, the transformation

$$\Psi(x_1,\ldots,x_n) = \left(G_1^{-1}(F_1(x_1)),\ldots,G_n^{-1}(F_1(x_n)) \right)$$

maps $\mathbf{X}$ onto $\mathbf{Y}$ and, clearly, then $\Psi = \mathbf{Q_Y}(\mathbf{D_X})$ (see Section 1.3). Obviously, Ψ is dispersive if, and only if, $X_i \leq_{\text{disp}} Y_i$, for all $i = 1,\ldots,n$.   □

Again, if we consider random vectors with the same copula, whatever it would be, and given univariate marginal distributions ordered in the dispersive order, we can provide many examples of random vectors ordered in the multivariate dispersive order. Examples of univariate distributions ordered in the dispersive order can be found in Tables 2.1 and 2.3.

Let us see another interesting situation where the comparison of the marginal distributions in the dispersive order becomes the key to obtain different stochastic comparisons.

Suppose that two random vectors have a common copula with the PA property (see Section 1.3). In this case, the comparison in the univariate dispersive order of the marginal distributions is a sufficient condition for the comparison in a particular stochastic order of increasing directionally convex transformations of the random vectors.

Let us recall the definition of an univariate stochastic order that compares the variances, denoted by Var, of certain functions rather than just the expected values.

Definition 3.6.5. Given two random variables X and Y, such that $X \leq_{\text{st}} Y$, we say that X is smaller than Y in the *st:icx order*, denoted by $X \leq_{\text{st:icx}} Y$, if

$$\text{Var}[h(X)] \leq \text{Var}[h(Y)],$$

for all increasing convex functions h, such that the variances exist.

As we have seen in Chapter 2, the st and icx orders do not guarantee the order among the variances. Thus, the st:icx order is useful in applications where a mere inequality between expected values is not sufficient for optimality evaluations.

Theorem 3.6.6. *Let* $\mathbf{X} = (X_1,\ldots,X_n)$ *and* $\mathbf{Y} = (Y_1,\ldots,Y_n)$ *be two continuous random vectors with a common PA copula C. If* $X_i \leq_{\text{st}} Y_i$, *for all* $i = 1,\ldots,n$, *then*

$$\phi(\mathbf{X}) \leq_{\text{st:icx}} \phi(\mathbf{Y}),$$

for all increasing directionally convex function $\phi : \mathbb{R}^n \mapsto \mathbb{R}$.

Proof. Let $\phi : \mathbb{R}^n \mapsto R$ be an idcx transformation. It is known that, given two random vectors with a common copula, if the marginal distributions are ordered in the stochastic order, that is, if $X_i \leq_{\text{st}} Y_i$, for all $i = 1, \ldots, n$, then $\mathbf{X} \leq_{\text{st}} \mathbf{Y}$. Since ϕ is increasing, $\phi(\mathbf{X}) \leq_{\text{st}} \phi(\mathbf{Y})$ holds.

Next, it is proved that the variances of $\phi(\mathbf{X})$ and $\phi(\mathbf{Y})$ are ordered. Let F_i and G_i the distribution functions of the marginal distributions of $\mathbf{X}$ and $\mathbf{Y}$, respectively. If $\mathbf{U} = (U_1, \ldots, U_n)$ is a random vector with joint distribution C, then we see that

$$\mathbf{X} =_{\text{st}} \mathbf{X}' = (F_1^{-1}(U_1), \ldots, F_n^{-1}(U_n)),$$
$$\mathbf{Y} =_{\text{st}} \mathbf{Y}' = (G_1^{-1}(U_1), \ldots, G_n^{-1}(U_n)),$$

and, additionally,

$$\mathbf{U} =_{\text{st}} (F_1(X_1), \ldots, F_n(X_n)) =_{\text{st}} (G_1(Y_1), \ldots G_n(Y_n)).$$

Since $X_i \leq_{\text{disp}} Y_i$, for all $i = 1, \ldots, n$, from Definition 2.6.1, we see that

$$(G_1^{-1}(p_1), \ldots, G_1^{-1}(p_n)) - (F_1^{-1}(p_1), \ldots, F_1^{-1}(p_n)) \text{ increases in } p_i.$$
(3.16)

From $X_i \leq_{\text{st}} Y_i$, for all $i = 1, \ldots, n$, and Definition 2.2.1, we find

$$(F_1^{-1}(p_1), \ldots, F_1^{-1}(p_n)) \leq (G_1^{-1}(p_1), \ldots, G_n^{-1}(p_n)),$$

for all $p_i \in (0, 1)$ and $i = 1, \ldots, n$.

According to the previous notation, let us prove that $\text{Var}[\phi(\mathbf{X}')] \leq \text{Var}[\phi(\mathbf{Y}')]$. Taking into account that

$$E[\phi^2(\mathbf{Y}')] - E[\phi(\mathbf{X}')\phi(\mathbf{Y}')] = E[\phi(\mathbf{Y}')(\phi(\mathbf{Y}') - \phi(\mathbf{X}'))],$$

we get $E[\phi^2(\mathbf{Y}')] - E[\phi(\mathbf{X}')\phi(\mathbf{Y}')]$, which can be written as

$$E[\phi^2(\mathbf{Y}')] - E[\phi(\mathbf{X}')\phi(\mathbf{Y}')] = E[H_1(U_1, \ldots, U_n)H_2(U_1, \ldots, U_n)],$$

where

$$H_1(p_1, \ldots, p_n) = \phi(G_1^{-1}(p_1), \ldots, G_n^{-1}(p_n))$$

and

$$H_2(p_1, \ldots, p_n) = \phi(G_1^{-1}(p_1), \ldots, G_n^{-1}(p_n)) - \phi(F_1^{-1}(p_1), \ldots, F_n^{-1}(p_n)).$$

It is clear that H_1 is an increasing function. In addition, combining (3.16), the increasing property of the univariate quantile function and the increasing directional convexity of ϕ, it follows that H_2 is also increasing. Therefore, due to the fact that $(U_1, \ldots, U_n)$ is PA, we see that

$$
\begin{aligned}
E[\phi^2(\mathbf{Y}')] - E[\phi(\mathbf{X}')\phi(\mathbf{Y}')] &= E[H_1(U_1, \ldots, U_n)H_2(U_1, \ldots, U_n)] \\
&\geq E[H_1(U_1, \ldots, U_n)]E[H_2(U_1, \ldots, U_n)] \\
&= (E[\phi(\mathbf{Y}')])^2 - E[\phi(\mathbf{X}')]E[\phi(\mathbf{Y}')].
\end{aligned}
$$

Thus, we get

$$
\text{Var}[\phi(\mathbf{Y})] \geq E[\phi(\mathbf{X}')\phi(\mathbf{Y}')] - E[\phi(\mathbf{X}')]E[\phi(\mathbf{Y}')].
$$

In a similar way, if we consider $E[\phi(\mathbf{X}')(\phi(\mathbf{Y}') - \phi(\mathbf{X}'))]$, it can be shown that

$$
\text{Var}[\phi(\mathbf{X})] \leq E[\phi(\mathbf{X}')\phi(\mathbf{Y}')] - E[\phi(\mathbf{X}')]E[\phi(\mathbf{Y}')].
$$

Therefore, $\text{Var}[\phi(\mathbf{X})] \leq \text{Var}[\phi(\mathbf{Y})]$, for all increasing directionally convex function ϕ.

Now, let $h : \mathbb{R} \mapsto \mathbb{R}$ be an increasing convex function and let ϕ be an increasing directionally convex function. The composition $h(\phi)$ is then increasing directionally convex (see Corollary 2.5 in Ref. [125]). Consequently, $\text{Var}[h(\phi(\mathbf{X}))] \leq \text{Var}[h(\phi(\mathbf{Y}))]$ for all increasing convex function h and any increasing directionally convex function ϕ. This completes the proof. □

As a particular case, the following result is provided.

Corollary 3.6.7. *Let* $\mathbf{X} = (X_1, \ldots, X_n)$ *and* $\mathbf{Y} = (Y_1, \ldots, Y_n)$ *be two continuous random vectors with a common PA copula C. Then,*

$$
\text{Var}\left[h\left(\sum_{i=1}^n a_i X_i \right) \right] \leq \text{Var}\left[h\left(\sum_{i=1}^n a_i Y_i \right) \right],
$$

for all increasing convex function h and $a_1 \ldots, a_n \geq 0$.

Proof. The result follows immediately, taking into account that

$$
\phi(x_1, \ldots, x_n) = h\left(\sum_{i=1}^n a_i X_i \right)
$$

is an increasing directionally convex function. □

Note that the assumptions in Theorem 3.6.6 are not too unrealistic. Let us consider two random vectors $\mathbf{X}$ and $\mathbf{Y}$, where $\mathbf{Y}$ has been obtained through a scale change of $\mathbf{X}$. Therefore, it is known that they have the same copula. For example, in risk theory, very often it is assumed that the individual risks are independent but this assumption can lead to a dramatic error [3]. Therefore, for real data, the individual risks could have some positive correlation with each other. Moreover, if we take into account the relationships between the univariate stochastic orders, the same result as in Theorem 3.6.6 is obtained but just considering $X_i \leq_{\text{disp}} Y_i$, where the left endpoints of the support of X_i and Y_i are equal, for all $i = 1, \ldots, n$. For example, this is the case of individual risks for non-negative random variables.

Finally, Shaked and Shanthikumar [72] provides a similar result in this direction, although under different assumptions. Let us give the result. The proof is omitted and can be seen in Ref. [72].

Theorem 3.6.8. *Let* $\mathbf{X} = (X_1, \ldots, X_n)$ *and* $\mathbf{Y} = (Y_1, \ldots, Y_n)$ *be two CIS continuous non-negative random vectors. If*

$$\mathbf{Q}_\mathbf{Y}(\mathbf{p}) - \mathbf{Q}_\mathbf{X}(\mathbf{p}) \text{ is increasing in } \mathbf{p} \in (0, 1)^n, \qquad (3.17)$$

then

$$\phi(\mathbf{X}) \leq_{\text{st:icx}} \phi(\mathbf{Y}),$$

for all increasing directionally convex function $\phi : \mathbb{R}^n \mapsto \mathbb{R}$.

Observe that (3.17) is sometimes considered as a generalization of the univariate dispersive order.

3.7 APPLICATIONS

In this section, applications on the comparison of mixtures of conditionally independent random variables in reliability and risks are described, as well as several results for the comparison of ordered data.

3.7.1 Comparisons of mixtures of conditionally independent random vectors

In previous sections, several results for the comparison of mixtures of conditionally random vectors are provided. Some additional results on this topic can be found in Refs. [126, 127]. Next, two applications of this model in the contexts of reliability and risks are described.

A useful model in reliability and survival analysis to model dependence between components or individuals is the *frailty model*. Let us describe the bivariate case. Given a bivariate random vector $\mathbf{X} = (X_1, X_2)$, we say that $\mathbf{X}$ follows a *bivariate correlated frailty model* if its joint survival function is given by

$$\overline{F}(x_1, x_2) = \int \overline{F}_1^{\theta_1}(x_1) \overline{F}_2^{\theta_2}(x_2) \, dH(\theta_1, \theta_2),$$

where H is the distribution of a bivariate random vector $\mathbf{\Theta} = (\Theta_1, \Theta_2)$ and $\overline{F}_i$ is a survival function, for $i = 1, 2$. Usually, $\mathbf{\Theta}$ is called the frailty random vector, which describes common random risk factors, and $\overline{F}_i$ is called the baseline survival function of X_i, for $i = 1, 2$. Notice that the components X_1 and X_2 are independent given any realization of the frailty random vector. When the random risk factors are common to both components—that is, if $\mathbf{\Theta} = (\Theta, \Theta)$ (and, therefore, H is a univariate distribution), then it is called a *bivariate shared frailty model*.

Next, some applications on previous results for conditionally random vectors to compare two frailty models are given.

Let $\mathbf{X} = (X_1, X_2)$ a bivariate correlated frailty model with frailty random vector $\mathbf{\Theta}_1$ and baseline survival functions $\overline{F}_i$, for $i = 1, 2$, and $\mathbf{Y} = (Y_1, Y_2)$ another bivariate correlated frailty model with frailty random vector $\mathbf{\Theta}_2$ and baseline survival functions $\overline{F}_i$, for $i = 1, 2$. Let us denote $X_i(\theta) = [X_i | \mathbf{\Theta}_{1i} = \theta]$ and $Y_i(\theta) = [X_i | \mathbf{\Theta}_{2i} = \theta]$, for $i = 1, 2$. It is then easy to see that $X_i(\theta)$ and $Y_i(\theta)$ are decreasing in the stochastic order in θ, for $i = 1, 2$. Obviously, $X_i(\theta) =_{st} Y_i(\theta)$, for all θ and $i = 1, 2$, therefore, from Theorem 3.2.10, if $\mathbf{\Theta}_1 \leq_{st} \mathbf{\Theta}_2$, then $\mathbf{X} \geq_{st} \mathbf{Y}$. Additional results for this model can be found in Refs. [112, 128, 129].

Another important application of mixtures of conditionally independent random vectors is given in portfolio credit risk. In this context, if we consider a portfolio with respect to n different obligors, the default risks of each obligor is assumed to depend on a set of economic factors, which are modelled stochastically. Given a realization of the factors, defaults of individual firms are assumed to be independent. Let us examine a particular case.

Let us consider the case where the default probability of the ith firm, depends on some random economic factors $\mathbf{\Theta} = \theta$, and it is denoted by $p_i(\theta)$. If $X_i(\theta)$ denotes the indicator random variable of default of ith firm, then $X_i(\boldsymbol{\theta})$ is a Bernoulli random variable with parameter $p_i(\boldsymbol{\theta}) = P[X_i(\boldsymbol{\theta}) = 1]$. If we consider the unconditional distribution of defaults of the n firms, $(X_1, \ldots, X_n)$, obtained by integrating over the distribution of the economic factors $\boldsymbol{\Theta}$, then $(X_1, \ldots, X_n)$ is said to follow a *Bernoulli mixture model* [50, p. 219].

Observe that, given a function $\phi : \mathbb{R}^n \mapsto \mathbb{R}$, we see that

$$E[\phi(X_i(\theta))] = \phi(0) + (\phi(1) - \phi(0))p_i(\boldsymbol{\theta}). \qquad (3.18)$$

Therefore, if ϕ is increasing, then the behaviour of $E[\phi(X_i(\theta))]$, with respect to θ depends only on $p_i(\boldsymbol{\theta})$. For example, if $p_i(\boldsymbol{\theta})$ is increasing in θ, then $E[\phi(X_i(\theta))]$ is increasing in θ. Let us consider another set of n Bernoulli random variables, $Y_1(\boldsymbol{\theta}), \ldots, Y_n(\boldsymbol{\theta})$, where $P[Y_i(\boldsymbol{\theta}) = 1] = q_i(\boldsymbol{\theta})$, that can be considered as the random defaults of another set of firms given a realization of the economic factors, $\boldsymbol{\Theta} = \boldsymbol{\theta}$. It is not difficult to see that, if $p_i(\boldsymbol{\theta}) \le q_i(\boldsymbol{\theta})$ then $X_i(\boldsymbol{\theta}) \le_{\text{st}} Y_i(\boldsymbol{\theta})$. Therefore, as a consequence of Theorem 3.2.10, $(X_1, \ldots, X_n) \le_{\text{st}} (Y_1, \ldots, Y_n)$. It can be also considered the situation where the default probability of ith firm can be computed under two different scenarios, $\boldsymbol{\Theta}_1$ and $\boldsymbol{\Theta}_2$, as $P[\text{default of }i\text{th firm}|\boldsymbol{\Theta}_1 = \boldsymbol{\theta}] = p_i(\boldsymbol{\theta})$ and $P[\text{default of }i\text{th firm}|\boldsymbol{\Theta}_2 = \boldsymbol{\theta}] = q_i(\boldsymbol{\theta})$, respectively. From the previous conditions and assuming that $\boldsymbol{\Theta}_1 \le_{\text{st}} \boldsymbol{\Theta}_2$, we have again, from Theorem 3.2.10, that $(X_1, \ldots, X_n) \le_{\text{st}} (Y_1, \ldots, Y_n)$. Let us consider now the potential loss of ith firm, given by e_i, where e_i is positive and deterministic. The portfolio loss under the two scenarios are then given by $L_1 = \sum_{i=1}^{n} e_i X_i$ and $L_2 = \sum_{i=1}^{n} e_i Y_i$, and under previous considerations we obtain that $L_1 \le_{\text{st}} L_2$.

Now, let us consider, for example, that the two scenarios are modelled by two multivariate logit-normal distributions—that is, for $i = 1, 2$

$$\boldsymbol{\Theta}_i = \left(\frac{\exp\{Z_1^i\}}{1 + \exp\{Z_1^i\}}, \ldots, \frac{\exp\{Z_m^i\}}{1 + \exp\{Z_m^i\}} \right),$$

where $\boldsymbol{Z}_1 \sim N_n(\boldsymbol{\mu}_1, \boldsymbol{\Sigma}_1)$ and $\boldsymbol{Z}_2 \sim N_n(\boldsymbol{\mu}_2, \boldsymbol{\Sigma}_2)$. Given that the multivariate stochastic order is preserved under increasing transformations, and from Example 3.2.7, if $\boldsymbol{\mu}_1 \le \boldsymbol{\mu}_2$ and $\boldsymbol{\Sigma}_1 = \boldsymbol{\Sigma}_2$, then $\boldsymbol{\Theta}_1 \le_{\text{st}} \boldsymbol{\Theta}_2$. Further results in this context can be found in Ref. [112].

3.7.2 Comparisons of ordered data
Next, some results for the comparison of ordered data are given. Examples of ordered data are the usual order statistics and record values. Given the

similarity of several results for order statistics and record values, Kamps [130] introduces the model of generalized order statistics. This model provides a unified approach to study order statistics and record values at the same time, among several other models of ordered data.

First, the definition of the generalized order statistics is recalled [130, 131].

Definition 3.7.1. Given the parameters $n \in \mathbb{N}$, $k \geq 1$, $m_1, \ldots, m_{n-1} \in \mathbb{R}$ and $M_r = \sum_{j=r}^{n-1} m_j$, for all $1 \leq r \leq n-1$, such that $\gamma_r = k+n-r+M_r \geq 1$, for all $r \in 1, \ldots, n-1$, and let $\tilde{m} = (m_1, \ldots, m_{n-1})$, if $n \geq 2$ ($\tilde{m} \in \mathbb{R}$ arbitrary, if $n = 1$), it is said that $(U_{(1,n,\tilde{m},k)}, \ldots, U_{(n,n,\tilde{m},k)})$ is the random vector of *uniform generalized order statistics*, if its joint density function is given by

$$h(u_1, \ldots, u_n) = k \left(\prod_{j=1}^{n-1} \gamma_j \right) \left(\prod_{j=1}^{n-1} (1 - u_j)^{m_j} \right) (1 - u_n)^{k-1},$$

for all $0 \leq u_1 \leq \ldots \leq u_n \leq 1$. Now, given a distribution function F, it is said that

$$(X_{(1,n,\tilde{m},k)}, \ldots, X_{(n,n,\tilde{m},k)}) = \left(F^{-1}(U_{(1,n,\tilde{m},k)}), \ldots, F^{-1}(U_{(n,n,\tilde{m},k)}) \right),$$

is the random vector of *generalized order statistics* (GOS) based on F.

If F is a differentiable distribution function with density function f, the joint density function of $(X_{(1,n,\tilde{m},k)}, \ldots, X_{(n,n,\tilde{m},k)})$ is given by

$$f(x_1, \ldots, x_n) = k \left(\prod_{j=1}^{n-1} \gamma_j \right) \left(\prod_{j=1}^{n-1} \bar{F}(x_j)^{m_j} f(x_j) \right) \bar{F}(x_n)^{k-1} f(x_n),$$

for all $F^{-1}(0) \leq x_1 \leq \cdots \leq x_n \leq F^{-1}(1)$.

Among the different distributional properties of GOS, we recall that two random vectors of GOS with the same set of parameters and possibly based on different distributions have the same copula. Another interesting property is that, according to the notation by Cramer and Kamps [132], the survival function of $X_{(r,n,\tilde{m}_n,k)}$ can be written as

$$\bar{F}_{*,r} = H_r(\bar{F}(t)), \quad \text{for all } t \in \mathbb{R}, \tag{3.19}$$

where H_r is defined by

$$H_r(z) = \left(\prod_{j=1}^{r} \gamma_j\right) \int_0^z G_{r,r}^{r,0}\left[s \left|\begin{array}{c} \gamma_1,\ldots,\gamma_r \\ \gamma_1 - 1,\ldots,\gamma_r - 1 \end{array}\right.\right] ds, \text{ for } z \in (0,1),$$

and G is a particular Meijer's G-function.

As we have mentioned previously, the order statistics and record values are particular cases of this model. On the one hand, taking $m_i = 0$, for all $i = 1,\ldots,n - 1$ and $k = 1$, the order statistics from a distribution F arises. On the other hand, taking $m_i = -1$ for all $i = 1,\ldots,n - 1$ and $k = 1$ we get the first n record values from a sequence of random variables with distribution F. Other particular cases are k-record values, progressively type II censored order statistics and order statistics under multivariate imperfect repair (see Ref. [133] for details).

Due to the fact that two random vectors of GOS, with the same set of parameters, have the same copula, we can apply previous results to provide multivariate comparisons of random vectors of GOS with the same set of parameters. Let us describe the situation.

Let X and Y be random variables with distribution functions F and G respectively, and let $(X_{(1,n,\tilde{m},k)},\ldots,X_{(n,n,\tilde{m},k)})$ and $(Y_{(1,n,\tilde{m},k)},\ldots,Y_{(n,n,\tilde{m},k)})$ be two random vectors of generalized order statistics based on F and G, respectively. From Theorems 3.2.6, 3.3.6 and 3.6.4, we see that:

(i) If $X_{(i,n,\tilde{m},k)} \leq_{st} Y_{(i,n,\tilde{m},k)}$, for all $i = 1,\ldots,n$, then

$$(X_{(1,n,\tilde{m},k)},\ldots,X_{(n,n,\tilde{m},k)}) \leq_{st} (Y_{(1,n,\tilde{m},k)},\ldots,Y_{(n,n,\tilde{m},k)}).$$

(ii) If $X_{(i,n,\tilde{m},k)} \leq_{icx} Y_{(i,n,\tilde{m},k)}$, for all $i = 1,\ldots,n$, then

$$(X_{(1,n,\tilde{m},k)},\ldots,X_{(n,n,\tilde{m},k)}) \leq_{idcx} (Y_{(1,n,\tilde{m},k)},\ldots,Y_{(n,n,\tilde{m},k)}).$$

(iii) If $X_{(i,n,\tilde{m},k)} \leq_{disp} Y_{(i,n,\tilde{m},k)}$, for all $i = 1,\ldots,n$, then

$$(X_{(1,n,\tilde{m},k)},\ldots,X_{(n,n,\tilde{m},k)}) \leq_{disp} (Y_{(1,n,\tilde{m},k)},\ldots,Y_{(n,n,\tilde{m},k)}).$$

The result in (b) also requires the copula to be CI, but this follows easily, since the copula of GOS is always MTP$_2$.

The question now is to compare the components in the univariate stochastic, increasing convex and dispersive orders. From (3.19), it is obvious that $G_{*,r}^{-1}(F_{*,r}) = G^{-1}(F)$ and, therefore, $X_{(i,n,\tilde{m},k)} \leq_{st[disp]} Y_{(i,n,\tilde{m},k)}$, for all $i = 1,\ldots,n$, if, and only if, $X \leq_{st[disp]} Y$.

There is no equivalent result for the icx order. In any case, Balakrishnan et al. [115] prove that if $X_{(1,n,\tilde{m},k)} \leq_{icx} Y_{(1,n,\tilde{m},k)}$, then $X_{(i,n,\tilde{m},k)} \leq_{icx} Y_{(i,n,\tilde{m},k)}$, for all $i = 1, \ldots, n$. Hence, it is enough to check if the minimums are ordered in the icx order to compare the whole random vectors in the idcx order.

Let us recall now a result for the multivariate likelihood ratio order. For a detailed proof the reader can look in Ref. [134].

Let X and Y be two random variables with differentiable distribution functions and hazard rates functions r and s, respectively. Let $\mathbf{X} = (X_{(1,n,\tilde{m},k)}, \ldots, X_{(n,n,\tilde{m},k)})$ and $\mathbf{Y} = (Y_{(1,n,\tilde{m},k)}, \ldots, Y_{(n,n,\tilde{m},k)})$ be two random vectors of generalized order statistics based on F and G, respectively. If one of the following sets of conditions holds:

(i) $m_i \geq 0$ for all i, and $X \leq_{lr} Y$, or
(ii) $m_i \geq -1$ for all i, $X \leq_{hr} Y$ and s/r is increasing,

then

$$(X_{(1,n,\tilde{m},k)}, \ldots, X_{(n,n,\tilde{m},k)}) \leq_{lr} (Y_{(1,n,\tilde{m},k)}, \ldots, Y_{(n,n,\tilde{m},k)}).$$

Given that the multivariate likelihood ratio order is preserved under marginalization, under previous conditions, we also have that the marginal distributions are also ordered in the univariate likelihood ratio order.

Let us see now how two random vectors of GOS can be compared in the multivariate dynamic hazard rate order.

First, an explicit expression for the multivariate dynamic hazard rates of random vectors of GOS is given. Let us denote by $\eta.(\cdot|\cdot)$ and $\lambda.(\cdot|\cdot)$ the multivariate dynamic hazard rate functions associated with $(X_{(1,n,\tilde{m},k)}, \ldots, X_{(n,n,\tilde{m},k)})$ and $(Y_{(1,n,\tilde{m},k)}, \ldots, Y_{(n,n,\tilde{m},k)})$, respectively, and let us consider two histories h_t and h'_t as in Definition 3.4.1, respectively. Since $X_{(1,n,\tilde{m},k)} \leq \cdots \leq X_{(n,n,\tilde{m},k)}$, a.s., and analogously for the random vectors of GOS based on G, we see that $I = \{1, \ldots, l\}$, $J = \{l+1, \ldots, m\}$. From the Markovian property of the GOS, we see that

$$\eta_k(t|h_t) = \begin{cases} r(t) & \text{if } k = m+1 \\ 0 & \text{if } k > m+1, \end{cases}$$

and

$$\lambda_k(t|h'_t) = \begin{cases} s(t) & \text{if } k = l+1 \\ 0 & \text{if } k > l+1, \end{cases}$$

where r and s denote the hazard rates of X and Y, respectively.

Suppose now that $X \leq_{\mathrm{hr}} Y$. Given $k \in \overline{I \cup J}$, that is, $k > l$, we see that, if $m > l$, then

$$
\begin{aligned}
\eta_k(t\,|\,h_t) &= r(t) \geq 0 = \lambda_k(t\,|\,h'_t), &&\text{if } k = l+1, \\
\eta_k(t\,|\,h_t) &= 0 = \lambda_k(t\,|\,h'_t), &&\text{if } k > l+1
\end{aligned}
$$

If $l = m$, that is, $J = \emptyset$, then, if we assume $X \leq_{\mathrm{hr}} Y$, we get

$$
\begin{aligned}
\eta_k(t\,|\,h_t) &= r(t) \geq s(t) = \lambda_k(t\,|\,h'_t), &&\text{if } k = l+1, \\
\eta_k(t\,|\,h_t) &= 0 = \lambda_k(t\,|\,h'_t), &&\text{if } k > l+1
\end{aligned}
$$

Therefore,

$$
(X_{(1,n,\tilde{m},k)}, \ldots, X_{(n,n,\tilde{m},k)}) \leq_{\text{dyn-hr}} (Y_{(1,n,\tilde{m},k)}, \ldots, Y_{(n,n,\tilde{m},k)}).
$$

The multivariate dynamic hazard rate order is not known to be preserved under marginalization and, therefore, we cannot obtain from the previous result the hazard rate order of the corresponding coordinates. In any case, it is possible to prove that if $X \leq_{\mathrm{hr}} Y$, then $X_{(r,n,\tilde{m},k)} \leq_{\mathrm{hr}} Y_{(r,n,\tilde{m},k)}$.

This is just a sample of results for the comparison of ordered data. In fact, this topic is one of the most prolific fields where stochastic orders have been applied. Additional reviews and references on the topic can be found in Refs. [93, 94, 133, 135, 136].

3.8 SUMMARY

This chapter gives an introduction to the topic in the multivariate setting. The intention is to provide an easy first reading on multivariate stochastic orders. We have presented the most basic criteria to compare random vectors and several results on them, placing some emphasis on the application to the comparison of convolutions for possible dependent addends. Chapter 3 finishes with some applications to the comparison of frailty models, mixture models in credit risk and ordered data.

CONCLUSION

This book is an introduction to stochastic orders, aiming to provide a basic understanding on the comparison of random quantities for those who are dealing for the first time with the topic, in particular for graduate and PhD students who are starting their studies on stochastic orders, and researchers interested in the topic but who are not specialists on it. Stochastic orders allow us to compare random quantities in a more complete way when single measures do not take into account all the information. More specifically, they are based on the comparison of different functions that inform us about the location, dispersion, variability or concentration of the random phenomena, and are also meaningful in fields like reliability, survival analysis, risks, and economics.

In the literature, there are several books on stochastic orders and related notions; however, sometimes they are difficult to read for beginners, and the first contribution of this book is to give an easy-to-understand overview of this topic. For that reason, we have avoided any discussion that requires notions far beyond of basic probability theory and real analysis courses. It is important to notice that we have given motivations for the definition of the several stochastic orders not only from a mathematical point of view, but also from an applied one. The stochastic orders considered in this book have a direct application in reliability and survival analysis to compare lifetimes of units or individuals, in risk theory to compare portfolios and in economics to compare inequality among income populations.

As another contribution, the main results are illustrated with several examples. In particular, we have provided discussions on sufficient conditions for the different stochastic orders and applications to several parametric models. This is a novelty, and we consider that this will attract the attention of any non-expert reader. In Tables 2.1–2.5, we have collected the results described along the book and some additional results that can be found in the literature.

Moreover, we have included several sections on applications. Given that this book is an introduction, we have not tried to cover all the details but to give a glimpse. We have, however, given additional references for those interested in particular applications. In particular, for the univariate stochastic orders, we have included descriptions of comparisons of coherent systems in reliability and distorted distributions and individual and collective risks in risk theory. For the multivariate stochastic orders, we have provided descriptions of comparisons for the individual risk model with dependent risks, and frailty models in reliability. For both the univariate and the multivariate cases, we have provided some applications for the comparison of ordered data.

BIBLIOGRAPHY

[1] M. Shaked, J.G. Shanthikumar, Stochastic Orders and their Applications, Academic Press, San Diego, CA, 1994.

[2] M. Shaked, J.G. Shanthikumar, Stochastic Orders, Springer Series in Statistics, Springer, New York, 2007.

[3] A. Müller, D. Stoyan, Comparison Methods for Stochastic Models and Risks, John Wiley and Sons, Chichester, 2002.

[4] D. Stoyan, Comparison Methods for Queues and Other Stochastic Models, Wiley, New York, 1983.

[5] R. Kaas, A.E. van Heerwaarden, M.J. Goovaerts, Ordering of Actuarial Risks, Caire Education Series, Caire, Brussels, 1994.

[6] M. Denuit, J. Dhaene, M. Goovaerts, R. Kaas, Actuarial Theory for Dependent Risks, John Wiley and Sons, Chichester, England, 2005.

[7] H. Levy, Stochastic Dominance. Investment Decision Making under Unvertainty, Springer, New York, 2006.

[8] S. Sriboonchitta, W.K. Wong, S. Dhompongsa, H.T. Nguyen, Stochastic Dominance and Applications to Finance, Risk and Economics, CRC Press, Boca Raton, FL, 2010.

[9] R.J. Hyndman, Y. Fan, Sample quantiles in statistical packages, Amer. Statist. 50 (1996) 361–365.

[10] R.E. Barlow, F. Proschan, Statistical Theory of Reliability and Life Testing, second ed., Holt, Rinehart and Winston, Inc., New York, 1975.

[11] S. Karlin, F. Proschan, Pólya type distributions of convolutions, Ann. Math. Stat. 31 (1960) 721–736.

[12] B.C. Barlow, D.J. Bartholomew, J.M. Bremner, H.D. Brunk, Statistical Inference Under Order Restrictions, John Wiley and Sons, New York, 1972.

[13] F. Belzunce, J.F. Pinar, J.M. Ruíz, M.A. Sordo, Comparisons of risks based on the expected proportional shortfall, Insur. Math. Econ. 51 (2012) 292–302.

[14] W.G. Runciman, Relative Deprivation and Social Justice, Routledge and Kegan Paul, London, 1966.

[15] T. Hu, Y. Wang, W. Zhuang, New properties of the total time on test transform order, Probabil. Eng. Informat. Sci. 26 (2012) 43–60.

[16] S. Karlin, Total Positivity, Stanford University Press, Palo Alto, CA, 1968.

[17] K. Joagdev, S. Kochar, F. Proschan, A general composition theorem and its applications to certain partial orderings of distributions, Stat. Probabil. Lett. 22 (1995) 111–119.

[18] J.B. McDonald, Some generalized functions for the size distribution of income, Econometrica 52 (1984) 647–663.

[19] R. Bandourian, J.B. McDonald, R.S. Turley, A comparison of parametric models of income distribution across countries and over time, Estadística 55 (2004) 127–142.

[20] P.J. Boland, Statistical and Probabilistic Methods in Actuarial Science, Chapman and Hall, Boca Raton, FL, 2007.

[21] C. Kleiber, S. Kotz, Statistical Size Distributions in Economics and Actuarial Sciences, John Wiley and Sons, Hoboken, NJ, 2003.

[22] H.L. Seal, Survival probabilities based on Pareto claim distributions, ASTIN Bull. 11 (1980) 61–71.

[23] J. Beirlant, J.L. Teugels, Modeling large claims in non-life insurance, Insur. Math. Econ. 11 (1992) 17–29.

[24] R.K.S. Hankin, A. Lee, A new family of non-negative distributions, Aust. NZ J. Stat. 48 (2006) 67–78.

[25] N.U. Nair, B. Vineshkumar, L-moments of residual life, J. Stat. Plan. Infer. 140 (2010) 2168–2631.

[26] Z. Govindarajulu, A class of distributions useful in lifetesting and reliability with applications to nonparametric testing, in: C.P. Tsokos, I.N. Shimi (Eds.), Theory and Applications of Reliability, vol. 1, Academic Press, New York, 1977, pp. 109–130.

[27] N.U. Nair, P.G. Sankaran, B. Vineshkumar, Characterization of distributions by properties of truncated Gini index and mean difference, Metron 70 (2012) 173–191.

[28] E. Arjas, T. Lehtonen, Approximating many server queues by means of single server queues, Math. Operat. Res. 3 (1978) 205–223.

[29] G.L. O'Brien, The comparison method for stochastic processes, Ann. Probabil. 3 (1975) 80–88.

[30] M. Rosenblatt, Remarks on a multivariate transformation, Ann. Math. Stat. 23 (1952) 470–472.

[31] L. Rüschendorf, Stochastically ordered distributions and monotonicity of the OC-function of sequential probability ratio tests, Math. Operat. Stat. Series Stat. 12 (1981) 327–338.

[32] J.M. Fernández-Ponce, A. Suárez-Llorens, A multivariate dispersion ordering based on quantiles more widely separated, J. Multivariate Anal. 85 (2003) 40–53.

[33] A. Sklar, Fonctions de répartition à n dimensions et leurs marges, Publ. Inst. Stat. Univ. Paris 8 (1959) 229–231.

[34] G. Kimeldorf, A. Sampson, Uniform representations of bivariate distributions with fixed marginals, Commun. Stat. Theory Meth. 4 (1975) 617–627.

[35] P. Deheuvels, Caractérisation compelet des lois extrêmes multivariées et de la convergence des types extrêmes, Publ. Inst. Stat. Univ. Paris 23 (1978) 1–37.

[36] E.W. Frees, J. Carriere, E.A. Valdez, Annuity valuation with dependent mortality, Probabil. Eng. Informat. Sci. 63 (1996) 229–261.

[37] E.W. Frees, E.A. Valdez, Understanding relationships using copulas, N. Amer. Actuar. J. 2 (1998) 1–25.

[38] E.W. Frees, P. Wang, Credibility using copulas, N. Amer. Actuar. J. 9 (2005) 31–48.

[39] E. Bouyé, V. Durrleman, A. Bikeghbali, G. Riboulet, T. Roncalli, Copulas for Finance—A Reading Guide and Some Applications, Social Science Research Network, 2000, available at: URL http://ssrn.com/abstract=1032533.

[40] P. Embrechts, F. Lindskog, A. McNeil, Modelling dependence with copulas and applications to risk management, in: S.T. Rachev (Ed.), Handbook of Heavy Tailed Distribution in Finance, Elsevier, Amsterdam, 2003, pp. 329–384.

[41] U. Cherubini, E. Luciano, W. Vecchiato, Copula Methods in Finance, John Wiley and Sons, Chichester, England, 2004.

[42] W. Wang, M.T. Wells, Model selection and semiparametric inference for bivariate failure-time data, J. Amer. Stat. Assoc. 95 (2000) 62–72.

[43] G. Escarela, J.F. Carrière, Fitting competing risks with an assumed copula, Stat. Meth. Med. Res. 12 (2003) 333–349.

[44] J. Yan, Multivariate modeling with copulas and engineering applications, in: H. Pham (Ed.), Handbook in Engineering Statistics, Springer-Verlag, New York, 2006, pp. 973–990.

[45] C. Genest, A.C. Favre, Everything you always wanted to know about copula modeling but were afraid to ask, J. Hydrol. Eng. 12 (2007) 347–368.

[46] R.B. Nelsen, An Introduction to Copulas, Lectures Notes in Statistics, vol. 139, Springer-Verlag, New York, 1999.

[47] M. Shaked, J.G. Shanthikumar, Multivariate hazard rates and stochastic ordering, Adv. Appl. Probabil. 19 (1987) 123–137.

[48] M. Shaked, J.G. Shanthikumar, Dynamic multivariate mean residual life functions, J. Appl. Probabil. 28 (1991) 613–629.

[49] M. Shaked, J.G. Shanthikumar, Multivariate conditional hazard rate functions—an overview, Appl. Stochast. Models Busi. Indust. 2014, doi:10.1002/asmb.2020.

[50] H. Joe, Multivariate Models and Dependence Concepts, Chapman and Hall, Boston, 1997.

[51] J.D. Esary, F. Proschan, D.W. Walkup, Association of random variables, with applications, Ann. Math. Stat. 38 (1967) 1466–1474.

[52] A.K. Gupta, T. Vargas, T. Bodnar, Elliptically Contoured Models in Statistics and Portfolio Theory, Springer-Verlag, New York, 2013.

[53] A. Müller, Stochastic orders generated by integrals: A unified approach, Adv. Appl. Probabil. 29 (1997) 414–428.

[54] D.G. Hoel, A representation of mortality data by competing risks, Biometrics 28 (1972) 475–489.

[55] M.A. Sordo, H.M. Ramos, Characterizations of stochastic orders by L-functionals, Stat. Papers 48 (2007) 249–263.

[56] S. Karlin, A. Novikoff, Generalized convex inequalities, Pacific J. Math. 13 (1963) 1251–1279.

[57] W. Hurliman, On stop-loss order and the distortion pricing principle, ASTIN Bull. 28 (1998) 119–134.

[58] O. Hesselager, Order relations for some distributions, Insur. Math. Econ. 15 (1995) 129–134.

[59] F. Belzunce, M. Shaked, Stochastic comparisons of mixtures of convexly ordered distributions with applications in reliability theory, Stat. Probabil. Lett. 53 (2001) 363–372.

[60] R.L. Berger, D.D. Boss, F.M. Guess, Tests and confidence sets for comparing two mean residual life functions, Biometrics 44 (1988) 103–115.

[61] A.K. Nanda, M. Shaked, The hazard rate and the reversed hazard rate orders, with applications to order statistics, Ann. Inst. Stat. Math. 53 (2001) 853–864.

[62] F. Belzunce, C. Martínez-Riquelme, J.M. Ruíz, On sufficient conditions for mean residual life and related orders, Comput. Stat. Data Anal. 61 (2013) 199–210.

[63] C. Metzger, L. Rüschendorf, Conditional variability ordering of distributions, Ann. Operat. Res. 32 (1991) 127–140.

[64] T. Lewis, J.W. Thompson, Dispersive distributions, and the connection between dispersivity and strong unimodality, J. Appl. Probabil. 18 (1981) 76–90.

[65] J.M. Mu noz-Pérez, Dispersive ordering by the spread function, Stat. Probabil. Lett. 10 (1990) 407–410.

[66] J. Bartoszewicz, Dispersive ordering and monotone failure rate distributions, Adv. Appl. Probabil. 17 (1985) 472–474.

[67] I. Bagai, S.C. Kochar, On tail-ordering and comparison of failure rates, Commun. Stat. Theory Meth. 15 (1986) 1377–1388.

[68] M. Shaked, Dispersive ordering of distributions, J. Appl. Probabil. 19 (1982) 310–320.

[69] D.J. Saunders, Dispersive ordering of distributions, Adv. Appl. Probabil. 16 (1984) 693–694.

[70] J.M. Fernández-Ponce, S.C. Kochar, J. Muñoz-Pérez, Partial orderings of distributions based on right-spread functions, J. Appl. Probabil. 35 (1998) 221–228.

[71] J. Bartoszewicz, Characterizations of stochastic orders based on ratios of Laplace transforms, Stat. Probabil. Lett. 42 (1999) 207–212.

[72] M. Shaked, J.G. Shanthikumar, Two variability orders, Probabil. Eng. Informat. Sci. 12 (1998) 1–23.

[73] F. Belzunce, C. Martínez-Riquelme, J.M. Ruíz, M.A. Sordo, On the comparison of relative spacings with applications, Technical Report, Dpto. Estadística e Investigación Operativa, Universidad de Murcia, 2015.

[74] T. Hu, J. Chen, J. Yao, Preservation of the location independent risk order under convolution, Insur. Math. Econ. 38 (2006) 406–412.

[75] J.G. Kuczmarski, P.R. Rosenbaum, Quantile plots, partial orders, and financial risk, Amer. Stat. 53 (1999) 239–246.

[76] F. Belzunce, J.F. Pinar, J.M. Ruíz, M.A. Sordo, Comparisons of concentration for several families of income distributions, Stat. Probabil. Lett. 83 (2013) 1036–1045.

[77] B.C. Arnold, Majorization and the Lorenz Order: A Brief Introduction, Springer, Berlin, 1987.

[78] J. Beirlant, Y. Goegebeur, J. Teugels, J. Segers, Statistics of Extremes. Theory and Applications, John Wiley and Sons, Chichester, 2004.

[79] F. Belzunce, C. Martínez-Riquelme, J.M. Ruíz, A characterization and sufficient conditions for the total time on test transform order, TEST 23 (2014) 72–85.

[80] S.C. Kochar, X. Li, M. Shaked, The total time on test transform and the excess wealth stochastic orders of distributions, Adv. Appl. Probabil. 34 (2002) 826–845.

[81] B. Proschan, Theoretical explanation of observed decreasing failure rate, Technometrics 5 (1963) 373–383.

[82] B. Lisek, Comparability of special distributions, Statistics 9 (1978) 537–593.

[83] J.M. Taylor, Comparisons of certain distributions functions, Statistics 14 (1983) 307–408.

[84] A. Klenke, L. Mattner, Stochastic ordering of classical discrete distributions, Appl. Probabil. Trust 42 (2010) 392–410.

[85] J. Lynch, G. Mimmack, F. Proschan, Uniform stochastic orderings and total positivity, Can. J. Stat. 15 (1987) 63–69.

[86] A.K. Nanda, K. Jain, H. Singh, Preservation of some partial orderings under the formation of coherent systems, Stat. Probabil. Lett. 39 (1998) 123–131.

[87] J. Bartoszewicz, Dispersive ordering and the total time on test transformation, Stat. Probabil. Lett. 4 (19) 285–288.

[88] F. Belzunce, M. Franco, J.M. Ruíz, M.C. Ruíz, On partial orderings between coherent systems with different structures, Probabil. Eng. Informat. Sci. 15 (2001) 273–293.

[89] J. Navarro, R. Rubio, Comparison of coherent systems using stochastic precedence, Test 19 (2010) 469–486.

[90] J. Navarro, R. Rubio, A note on necessary and sufficient conditions for ordering properties of coherent systems with exchangeable components, Naval Res. Logist. 58 (2011) 478–489.

[91]   J. Navarro, R. Rubio, Comparison of coherent systems with non-identically distributed components, J. Stat. Plan. Infer. 142 (2012) 1310–1319.

[92]   J.D. Esary, F. Proschan, Coherent structures of non-identical components, Technometrics 5 (1963) 191–209.

[93]   P.J. Boland, M. Shaked, J.G. Shanthikumar, Stochastic ordering of order statistics, in: N. Balakrishnan, C.R. Rao (Eds.), Handbook of Statistics, vol. 16, North-Holland, Amsterdam, 1998, pp, 89–103.

[94]   N. Balakrishnan, P. Zhao, Ordering properties of order statistics from heterogeneous populations: A review with an emphasis on some recent developments, Probabil. Eng. Informat. Sci. 27 (2013) 403–443.

[95]   M.E. Yaari, The dual theory of choice under risk, Econometrica 55 (1987) 95–115.

[96]   S. Wang, Premium calculation by transforming the layer premium density, ASTIN Bull. 26 (1996) 71–92.

[97]   J. Navarro, Y. del Aguila, M.A. Sordo, A. Suárez-Llorens, Stochastic ordering properties for systems with dependent identically distributed components, Appl. Stochast. Models Business Indust. 29 (2013) 264–278.

[98]   E. di Bernardino, D. Rullière, Distortions and multivariate distribution functions and associated level curves: Applications in multivariate risk theory, Insur. Math. Econ. 53 (2013) 190–205.

[99]   P. Zhao, X. Li, Ordering properties of convolutions from heterogeneous populations: A review on some recent developments, Commun. Stat. Theory Meth. 43 (2014) 2260–2273.

[100] F. Belzunce, R.E. Lillo, J.M. Ruíz, M. Shaked, Stochastic comparisons of nonhomogeneous processes, Probabil. Eng. Informat. Sci. 15 (2001) 199–224.

[101] H. Singh, K. Jain, Preservation of some partial orderings under Poisson shock models, J. Appl. Probabil. 21 (1989) 713–716.

[102] F. Pellerey, Partial orderings under cumulative shock models, Adv. Appl. Probabil. 25 (1993) 939–946.

[103] E. Fagiuoli, F. Pellerey, New partial orderings and applications, Naval Res. Logist. 40 (1993) 829–842.

[104] E. Fagiuoli, F. Pellerey, Mean residual life and increasing convex comparison of shock models, Stat. Probabil. Lett. 20 (1994) 337–345.

[105] Y. Kebir, Order-preserving shock models, Probabil. Eng. Informat. Sci. 8 (1994) 125–134.

[106] M. Shaked, T. Wong, Preservation of stochastic orderings under random mapping by point processes, Probabil. Eng. Informat. Sci. 9 (1995) 563–580.

[107] M. Shaked, T. Wong, Stochastic orders based on ratios of Laplace transforms, J. Appl. Probabil. 34 (1997) 404–419.

[108] A. Müller, Stochastic ordering of multivariate normal distributions, Ann. Inst. Stat. Math. 53 (2001) 567–575.

[109] Y. Ding, X. Zhang, Some stochastic orders of Kotz-type distributions, Stat. Probabil. Lett. 69 (2004) 389–396.

[110] Z. Landsman, A. Tsanakas, Stochastic ordering of bivariate elliptical distributions, Stat. Probabil. Lett. 76 (2006) 488–494.

[111] T. Kamae, U. Krengel, G.L. O'Brien, Stochastic inequalities on partially ordered spaces, Ann. Probabil. 5 (1977) 899–912.

[112] F. Belzunce, J.A. Mercader, J.M. Ruíz, F. Spizzichino, Stochastic comparison of multivariate mixture models, J. Multivar. Anal. 100 (2009) 1657–1669.

[113] M. Marinacci, L. Montrucchio, Ultramodular functions, Math. Operat. Res. 30 (2005) 311–332.

[114] M. Denuit, A. Müller, Smooth generators of integral stochastic orders, Ann. Appl. Probabil. 12 (2002) 1174–1184.

[115] N. Balakrishnan, F. Belzunce, M.A. Sordo, A. Suárez-Llorens, Increasing directionally convex orderings of random vectors having the same copula, and their use in comparing ordered data, J. Multivar. Anal. 105 (2012) 45–54.

[116] A. Müller, M. Scarsini, Stochastic comparison of random vectors with a common copula, Math. Operat. Res. 26 (2001) 723–740.

[117] M. Denuit, E. Frostig, Comparison of dependence in factor models with applications to credit risk portfolios, Probabil. Eng. Informat. Sci. 22 (2008) 151–160.

[118] T. Hu, B.E. Khaledi, M. Shaked, Multivariate hazard rate orders, J. Multivari. Anal. 84 (2003) 173–189.

[119] M. Shaked, J.G. Shanthikumar, Multivariate conditional hazard rate and mean residual life functions and their applications, in: R. Barlow, C.A. Clarotti, F. Spizzichino (Eds.), Reliability and Decision Making, Chapman & Hall, London, 1993, pp. 137–155.

[120] M. Shaked, J.G. Shanthikumar, J.B. Valdez-Torres, Discrete probabilistic orderings in reliability theory, Stat. Sin. 4 (1994) 567–579.

[121] M. Shaked, J.G. Shanthikumar, J.B. Valdez-Torres, Discrete hazard rate functions, Comput. Operat. Res. 22 (1995) 391–402.

[122] M. Shaked, J.G. Shanthikumar, Dynamic conditional marginal distributions in reliability theory, J. Appl. Probabil. 30 (1993b) 421–428.

[123] A. Giovagnoli, H.P. Wynn, Multivariate dispersion orderings, J. Appl. Probabil. 22 (1995) 325–332.

[124] D.A. Harville, Matrix Algebra from a Statistician's Perspective, Springer-Verlag, New York, 1997.

[125] L.E. Meester, J.G. Shanthikumar, Regularity of stochastic processes: A theory of directional convexity, Probabil. Eng. Informat. Sci. 7 (1993) 343–360.

[126] B.E. Khaledi, M. Shaked, Stochastic comparisons of multivariate mixtures, J. Multivar. Anal. 101 (2010) 2486–2498.

[127] X. Ling, P. Zhao, On general multivariate mixture models, Commun. Stat. Theory Meth. 43 (2014) 4223–4240.

[128] N. Misra, N. Gupta, R.D. Gupta, Stochastic comparisons of multivariate frailty models, J. Stat. Plan. Infer. 139 (2009) 2084–2090.

[129] J. Mulero, F. Pellerey, R. Rodríguez-Griñolo, Negative aging and stochastic comparisons of residual lifetimes in multivariate frailty models, J. Stat. Plan. Infer. 140 (2010) 1594–1600.

[130] U. Kamps, A concept of generalized order statistics, J. Stat. Plan. Infer. 48 (1995a) 1–23.

[131] U. Kamps, A Concept of Generalized Order Statistics, B.G. Taubner, Stuttgart, 1995b.

[132] E. Cramer, U. Kamps, Marginal distributions of sequential and generalized order statistics, Metrika 58 (2003) 293–310.

[133] F. Belzunce, Multivariate comparisons of ordered data, in: H. Li, X. Li (Eds.), Stochastic Orders in Reliability and Risks, Springer, New York, 2013, pp. 83–102.

[134] F. Belzunce, J.A. Mercader, J.M. Ruíz, Stochastic comparisons of generalized order statistics, Probabil. Eng. Informat. Sci. 23 (2005) 99–120.

[135] P.J. Boland, T. Hu, M. Shaked, J.G. Shanthikumar, Stochastic ordering of order statistics II, in: M. Dror, P. L'Ecuyer, F. Szidarovszky (Eds.), Modelling Uncertainty: An Examination of Stochastic Theory, Methods and Applications, Kluwer, Boston, 2002, pp. 607–623.

[136] F. Belzunce, R. Lillo, J.M. Ruíz, M. Shaked, Stochastic ordering of record and inter-record values, in: M. Ahsanullah, M. Raqab (Eds.), Recent Developments in Ordered Random Variables, Nova Sci. Publ., New York, 2007, pp. 119–137.

Printed in the United States
By Bookmasters